"A must for everyone curious about life in outer space and survival on Earth . . . As you read on, bells ring, lights flash, and you know that Sagan is right on target . . . His warmth and wit come across, making it possible for the most uninformed reader to enjoy and learn."

—*Los Angeles Free Press*

"Carl Sagan is a scientist of quality who is also a writer of quality. THE COSMIC CONNECTION is a book that is very nearly perfect. [With] great intelligence, wit and insight, it is a success on every level."

—*Washington Post*

"A profound and vastly entertaining book."

—WGN Radio

"Carl Sagan is the spokesman for hip astrophysics."

—*Changes*

"In simple language and shining images, Carl Sagan tells you why, and how, to understand this fantasy-like field."

—*Learning*

"Delightful, witty, amusing, thought-provoking, pithy, evocative. An astronomical perspective on 20th-Century Earth . . . Ought to be read by high-school and college kids, college drop-outs, your nephews and nieces."

—*Science*

"THE COSMIC CONNECTION . . . puts it all together."

—*Chicago Tribune*

"A daring view of the Universe by the wittiest and most clear-thinking astronomer alive today."

—Isaac Asimov

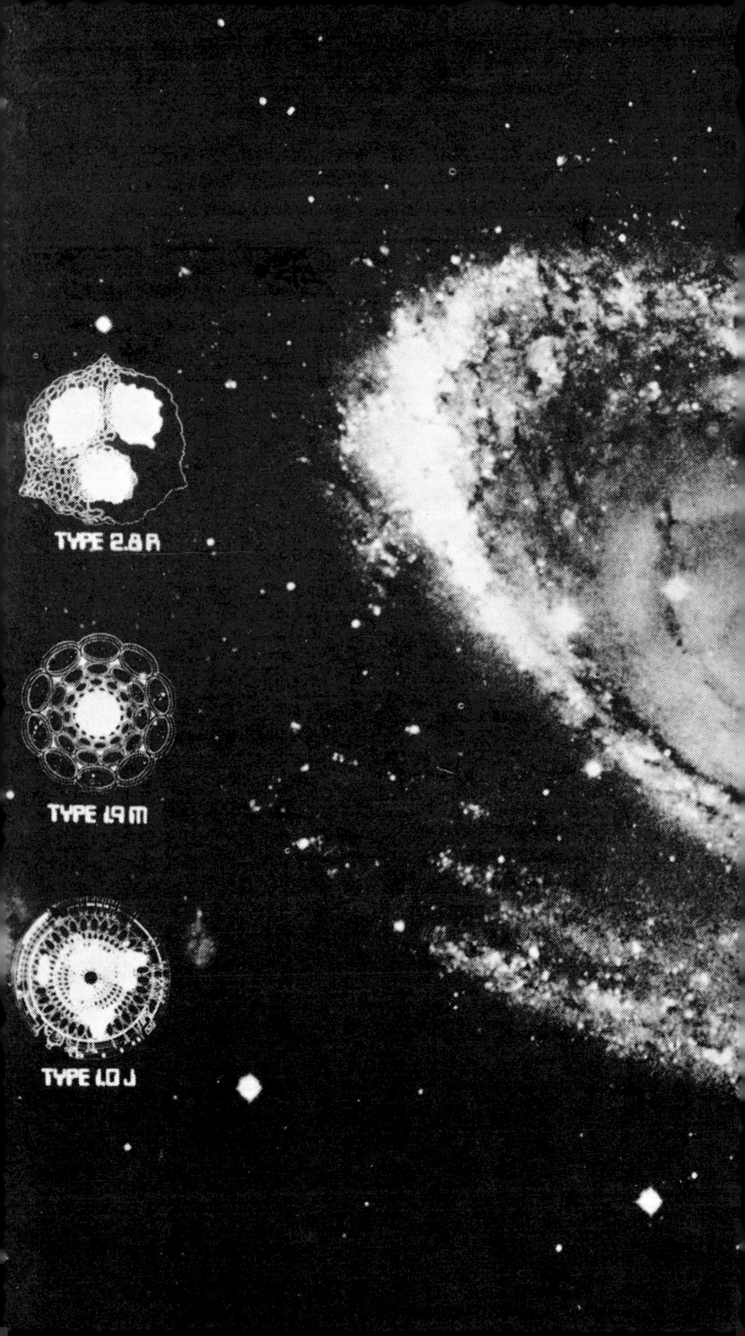
TYPE 2.8 R
TYPE 1.9 M
TYPE 1.0 J

THE COSMIC CONNECTION

An Extraterrestrial Perspective

by CARL SAGAN
Produced by JEROME AGEL

A DELL BOOK

Published by
DELL PUBLISHING CO., INC.
1 Dag Hammarskjold Plaza
New York, New York 10017

Dell ® TM 681510, Dell Publishing Co., Inc.
Reprinted by arrangement with
Doubleday & Company, Inc.
Printed in the United States of America
First Dell printing–March 1975
Second Dell printing–April 1975

DR. CARL SAGAN, the author of *The Cosmic Connection,* is Professor of Astronomy and Space Sciences and Director of the Laboratory for Planetary Studies at Cornell University. He received NASA's Medal for Exceptional Scientific Achievement for his studies of Mars with *Mariner 9*; he was responsible for placing the message from Earth aboard the interstellar spacecraft *Pioneer 10*; and he chaired the U.S. delegation to the U.S./U.S.S.R. Conference on Communication with Extraterrestrial Intelligence. Dr. Sagan was awarded in 1973 the Prix Galabert—the international astronautics prize. He is editor of the planetary science journal *Icarus* and is widely known for his studies of the planets, the origin of life, and the prospects for life beyond the Earth. He is formerly of the Harvard, Caltech and Stanford Medical School faculties.

JEROME AGEL is the producer of *The Cosmic Connection.* His fourteen book productions include: *Herman Kahnsciousness, Understanding Understanding* (with Humphrey Osmond), *The Medium Is the Massage* (with Marshall McLuhan), *The Making of Kubrick's "2001," Is Today Tomorrow?* (a synergistic collage of alternative futures), *I Seem to Be a Verb* (with Buckminster Fuller), *A World Without—What Our Presidents Didn't Know, Surprising Facts About U.S. History, Right on Time* (with Alan Lakein), *The Fasting Diet: How to Lose Weight Without Eating* (with Allan Cott), and *Rough Times.*

Illustration on title pages:
A spiral galaxy with representative Type I, Type II, and Type III civilizations indicated. By Jon Lomberg.

For Dorion, Jeremy, and Nicholas, my sons. May their future—and the future of all human and other beings—be bright with promise.

"Les Mystères des Infinis" by Grandville, 1844.

PREFACE

When I was twelve, my grandfather asked me—through a translator (he had never learned much English)—what I wanted to be when I grew up. I answered, "An astronomer," which, after a while, was also translated. "Yes," he replied, "but how will you make a living?"

I had supposed that, like all the adult men I knew, I would be consigned to a dull, repetitive, and uncreative job; astronomy would be done on weekends. It was not until my second year in high school that I discovered that some astronomers were paid to pursue their passion. I was overwhelmed with joy; I could pursue my interest full-time.

Even today, there are moments when what I do seems to me like an improbable, if unusually pleasant, dream: To be involved in the exploration of Venus, Mars, Jupiter, and Saturn; to try to duplicate the steps that led to the origin of life four billion years ago on an Earth very different from the one we know; to land instruments on Mars to search there for life; and perhaps to be engaged in a serious effort to communicate with other intelligent beings, if such there be, out there in the dark of the night sky.

Had I been born fifty years earlier, I could have pursued none of these activities. They were then all figments of the speculative imagination. Had I been born fifty years later, I also could not have been involved in these efforts, except possibly the last, because fifty years from now the preliminary reconnaissance of the Solar System, the search for life on Mars, and the study of the origin of life will have been completed. I think myself extraordinarily fortunate to be alive at the one moment in the history of mankind when such ventures are being undertaken.

So when Jerome Agel approached me about doing a popular book to try to communicate my sense of the excitement and importance of these adventures, I was amenable—even though his suggestion came just before the *Mariner* 9 mission to Mars, which I knew would occupy most of my waking hours for many months. At a later time, after discussing communication with extraterrestrial intelligence, Agel and I had dinner in a Polynesian restaurant in Boston. My fortune cookie announced, "You will shortly be called upon to decipher an important message." This seemed a good omen.

After centuries of muddy surmise, unfettered speculation, stodgy conservatism, and unimaginative disinterest, the subject of extraterrestrial life has finally come of age. It has now reached a practical stage where it can be pursued by rigorous scientific techniques, where it has achieved scientific respectability and where its significance is widely understood. Extraterrestrial life is an idea whose time has come.

This book is divided into three major sections. In the first part I try in several ways to convey a sense of cosmic perspective—living out our lives on a tiny hunk of rock and metal circling one of 250 billion stars that make up our galaxy in a universe of billions of galaxies. The deflation of some of our more common conceits is one of the practical applications of astronomy. The second part of the book is concerned with various aspects of our Solar System—mostly with Earth, Mars, and Venus. Some of the results and implications of *Mariner* 9 can be found here. Part Three is devoted to the possibility of communicating with extraterrestrial intelligence on planets of other stars. Since no such contact has yet been made—our efforts to date have been feeble—this section is necessarily speculative. I have not hesitated to speculate within what I perceive to be the bounds of scientific plausibility. And, although I am not by training a philosopher or sociologist or historian, I have not hesitated to draw philosophical or social or historical implications of astronomy and space exploration.

The astronomical discoveries we are in the midst of making are of the broadest human significance. If this book plays a small role in broadening public consideration of these exploratory ventures, it will have served its purpose.

As with all ongoing work and especially all speculative subjects, some of the statements in these pages will elicit vigorous demurrers. There are other books with other opinions. Reasoned disputation is the lifeblood of science—as is, sadly, infrequently the case in the intellectually more anemic arena of politics. But I believe that the more controversial opinions expressed here have, nevertheless, a significant scientific constituency. I have purposely introduced the same concept in slightly different contexts in a few places where I felt the discussion required it. The book is carefully structured, but, for the reader who wishes to browse ahead, most chapters are self-contained.

There are far too many who helped shape my opinions on these subjects for me to thank them all here. But in rereading these chapters, I find I owe a special debt to Joseph Veverka and Frank Drake, both of Cornell University, with whom over the past few years I have discussed so many aspects of this volume. The book was composed partly during a very long transcontinental trip in a very short automobile. I thank Linda and Nicholas for their encouragement and patience. I am also grateful to Linda for drawing two handsome humans and one elegant unicorn. And I am grateful to the late Mauritz Escher for permission to reproduce his "Another World" and to Robert Macintyre for the human figure and star field in Part Three. Jon Lomberg's paintings and drawings have been a source of intellectual and aesthetic excitement for me, and I am grateful to him for producing many of them especially for this book. Hermann Eckleman's careful photographic reproductions of Lomberg's work have facilitated their appearance in this book. And I thank Jerome Agel, without whose time and persistence this book would never have been written.

I am indebted to John Naugle of NASA for showing me his file on public response to the *Pioneer 10* plaque; the Oregon System of Higher Education for permission to reproduce some ideas from my book *Planetary Exploration;* the Forum for Contemporary History, in Santa Barbara, for permission to reproduce a portion of my letter distributed by the Forum in January 1973; and Cornell University Press for permission to reprint a fraction of my chapter "The Extraterrestrial and Other Hypotheses" from *UFO's: A Scientific Debate*, edited by Carl Sagan and Thornton Page, Cornell University Press, 1972. I am also grateful to those who have granted me permission to reproduce in Chapter 4 their remarks on the *Pioneer 10* plaque. The evolution of this book through many drafts owes much to the technical skills of Jo Ann Cowan, and, especially, Mary Szymanski.

—CARL SAGAN

CONTENTS

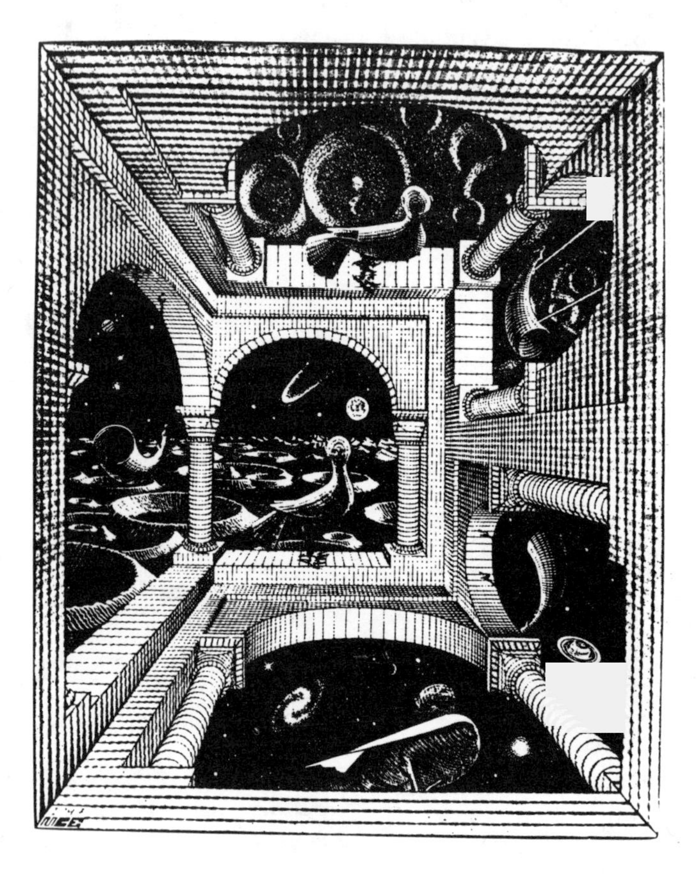

"Another World" by M. C. Escher.

Part One

COSMIC PERSPECTIVES

We shall not cease from exploration
And the end of all our exploring
Will be to arrive where we started
And know the place for the first time. . . .
When the tongues of flame are in-folded
Into the crowned knot of fire
And the fire and the rose are one.

—T. S. Eliot, *Four Quartets*

Untitled picture by Jon Lomberg.

1.
A Transitional Animal

Five billion years ago, when the Sun turned on, the Solar System was transformed from inky blackness to a flood of light. In the inner parts of the Solar System, the early planets were irregular collections of rock and metal—the debris, the minor constituents of the initial cloud, the material that had not been blown away after the Sun ignited.

These planets heated as they formed. Gases trapped in their interiors were exuded to form atmospheres. Their surfaces melted. Volcanoes were common.

The early atmospheres were composed of the most abundant atoms and were rich in hydrogen. Sunlight, falling on the molecules of the early atmosphere, excited them, induced molecular collisions, and produced larger molecules. Under the inexorable laws of chemistry and physics these molecules interacted, fell into the oceans, and further developed to produce larger molecules—molecules much more complex than the initial atoms of which they had formed, but still microscopic by any human standard.

These molecules, remarkably enough, are the ones of which we are made: The building blocks of the nucleic acids, which are our hereditary material, and the building blocks of the proteins, the molecular journeymen that perform the work of the cell, were produced from the atmosphere and oceans of the early Earth. We know this because we can make these molecules today by duplicating the primitive conditions.

Eventually, many billions of years ago, a molecule was formed that had a remarkable capability. It was able to produce, out of the molecular building blocks of the surrounding waters, a fairly accurate copy of itself. In such a molecular system there is a set of instructions, a molecular code, containing the sequence of building blocks from which the larger molecule is constructed. When, by accident, there is a change in the sequence, the copy is likewise changed. Such a molecular system—capable of replication, mutation, and replication of its mutations—can be called "alive." It is a collection of molecules that can evolve by natural selection. Those molecules able to replicate faster, or to reprocess building blocks from their surroundings into a more useful variety, reproduced more efficiently than their competitors—and eventually dominated.

But conditions gradually changed. Hydrogen escaped to space. Production of the molecular building blocks declined. The foodstuffs formerly available in great abundance dwindled. Life was expelled from the molecular Garden of Eden. Only those simple collections of molecules able to transform their surroundings, able to produce efficient molecular machines for the conversion of simple into complex molecules, were able to survive. By isolating themselves from their surroundings, by maintaining the earlier idyllic conditions, those molecules that surrounded themselves by membranes had an advantage. The first cells arose.

With molecular building blocks no longer available for free, organisms had to work hard to make such building blocks. Plants are the result. Plants start with air and water, minerals and sunlight, and produce molecular building blocks of high complexity. Animals, such as human beings, are parasites on the plants.

Changing climate and competition among what was now a wide diversity of organisms produced greater and greater specialization, a sophistication of function, and an elaboration of form. A rich array of plants and animals began to cover the Earth. Out of the initial oceans in which life arose, new environments, such as the land and the air, were colonized. Organisms now live from the top of Mount Everest to the deepest portions of the abyssal depths. Organisms live in hot, concentrated solutions of sulfuric acid and in dry Antarctic valleys. Organisms live on the water adsorbed on a single crystal of salt.

Life forms developed that were finely attuned to their specific environments, exquisitely adapted to the conditions. But the conditions changed. The organisms were too specialized. They died. Other organisms were less well adapted, but they were more generalized. The conditions changed, the climate varied, but the organisms were able to continue. Many more species of organisms have died during the history of the Earth than are alive today. The secret of evolution is time and death.

Among the adaptations that seem to be useful is one that we call intelligence. Intelligence is an extension of an evolutionary tendency apparent in the simplest organisms—the tendency toward control of the environment. The standby biological method of control has been the hereditary material: Information passed on by nucleic acids from generation to generation—information on how to build a nest; information on the fear of falling, or of snakes, or of the dark; information on how to fly south for the winter. But intelligence requires information of an adaptive quality developed during the lifetime of a single individual. A variety of organisms on the Earth today have this quality we call intelligence: The dolphins have it, and so do the great apes. But it is most evident in the organism called Man.

In Man, not only is adaptive information acquired in the lifetime of a single individual, but it is passed on extragenetically through learning, through books, through education. It is this, more than anything else, that has raised Man to his present pre-eminent status on the planet Earth.

We are the product of 4.5 billion years of fortuitous, slow, biological evolution. There is no reason to think that the evolutionary process has stopped. Man is a transitional animal. He is not the climax of creation.

The Earth and the Sun have life expectancies of many more billions of years. The future development of man will likely be a cooperative arrangement among controlled biological evolution, genetic engineering, and an intimate partnership between organisms and intelligent machines. But no one is in a position to make accurate predictions of this future evolution. All that is clear is that we cannot remain static.

In our earliest history, so far as we can tell, individuals held an allegiance toward their immediate tribal group, which may have numbered no more than ten or twenty individuals, all of whom were related by consanguinity. As time went on, the need for cooperative behavior—in the hunting of large animals or large herds, in agriculture, and in the development of cities—forced human beings into larger and larger groups. The group that was identified with, the tribal unit, enlarged at each stage of this evolution. Today, a particular instant in the 4.5-billion-year history of Earth and in the several-million-year history of mankind, most human beings owe their primary allegiance to the nation-state (although some of the most dangerous political problems still arise from tribal conflicts involving smaller population units).

Many visionary leaders have imagined a time when the allegiance of an individual human being is not to his particular nation-state, religion, race, or economic group, but to mankind as a whole; when the benefit to a human being of another sex, race, religion, or political persuasion ten thousand miles away is as precious to us as to our neighbor or our brother. The trend is in this direction, but it is agonizingly slow. There is a serious question whether such a global self-identification of mankind can be achieved before we destroy ourselves with the technological forces our intelligence has unleashed.

In a very real sense human beings are machines constructed by the nucleic acids to arrange for the efficient replication of more nucleic acids. In a sense our strongest urges, noblest enterprises, most compelling necessities, and apparent free wills are all an expression of the information coded in the genetic material: We are, in a way, temporary ambulatory repositories for our nucleic acids. This does not deny our humanity; it does not prevent us from pursuing the good, the true, and the beautiful. But it would be a great mistake to ignore where we have come from in our attempt to determine where we are going.

There is no doubt that our instinctual apparatus has changed little from the hunter-gatherer days of several hundred thousand years ago. Our society has changed enormously from those times, and the greatest problems of survival in the contemporary world can be understood in terms of this conflict—between what we feel we must do because of our primeval instincts and what we know we must do because of our extragenetic learning.

If we survive these perilous times, it is clear that even an identification with all of mankind is not the ultimate desirable identification. If we have a profound respect for other human beings as co-equal recipients of this precious patrimony of 4.5 billion years of evolution, why should the identification not apply also to all the other organisms on Earth, which are equally the product of 4.5 billion years of evolution? We care for a small fraction of the organisms on Earth—dogs, cats, and cows, for example—because they are useful or because they flatter us. But spiders and salamanders, salmon and sunflowers are equally our brothers and sisters.

I believe that the difficulty we all experience in extending our identification horizons in this way is itself genetic. Ants of one tribe will fight to the death intrusions by ants of another. Human history is filled with monstrous cases of small differences—in skin pigmentation, or abstruse theological speculation, or manner of dress and hair style—being the cause of harassment, enslavement, and murder.

A being quite like us, but with a small physiological difference—a third eye, say, or blue hair covering the nose and forehead—somehow evokes feelings of revulsion. Such feelings may have had adaptive value at one time in defending our small tribe against the beasts and neighbors. But in our times, such feelings are obsolete and dangerous.

The time has come for a respect, a reverence, not just for all human beings, but for all life forms—as we would have respect for a masterpiece of sculpture or an exquisitely tooled machine. This, of course, does not mean that we should abandon the imperatives for our own survival. Respect for the tetanus bacillus does not extend to volunteering our body as a culture medium. But at the same time we can recall that here is an organism with a biochemistry that tracks back deep into our planet's past. The tetanus bacillus is poisoned by molecular oxygen, which we breathe so freely. The tetanus bacillus, but not we, would be at home in the hydrogen-rich, oxygen-free atmosphere of primitive Earth.

A reverence for all life is implemented in a few of the religions of the planet Earth—for example, among the Jains of India. And something like this idea is responsible for vegetarianism, at least in the minds of many practitioners of this dietary constraint. But why is it better to kill plants than animals?

Human beings can survive only by killing other organisms. But we can make ecological compensation by also growing other organisms; by encouraging the forest; by preventing the wholesale slaughter of organisms such as seals and whales, imagined to have industrial or commercial value; by outlawing gratuitous hunting, and by making the environment of Earth more livable—for all its inhabitants.

There may be a time, as I describe in Part III of this book, when contact will be made with another intelligence on a planet of some far-distant star, beings with billions of years of quite independent evolution, beings with no prospect of looking very much like us—although they may think very much like us. It is important that we extend our identification horizons, not just down to the simplest and most humble forms of life on our own planet, but also up to the exotic and advanced forms of life that may inhabit, with us, our vast galaxy of stars.

2.
The Unicorn of Cetus

In the night sky, when the air is clear, there is a cosmic Rorschach test awaiting us. Thousands of stars, bright and faint, near and far, in a glittering variety of colors, are peppered across the canopy of night. The eye, irritated by randomness, seeking order, tends to organize into patterns these separate and distinct points of light. Our ancestors of thousands of years ago, who spent almost all their time out of doors in a pollution-free atmosphere, studied these patterns carefully. A rich mythological lore evolved.

Much of the original substance of this stellar mythology has not come down to us. It is so ancient, has been retold so many times, and especially in the past few thousand years by individuals unfamiliar with the appearance of the sky, that much has been lost. Here and there, in odd places, there remain some echoes of cosmic stories about patterns in the sky.

In the Book of Judges there is an account of a slain lion discovered to be infested by a hive of bees, a strange and apparently pointless incident. But the constellation of Leo in the night sky is adjacent to a cluster of stars, visible on a clear night as a fuzzy patch of light, called Praesepe. From its telescopic appearance, modern astronomers call it "The Beehive." I wonder if an image of Praesepe, obtained by one man of exceptional eyesight, in days before the telescope, has been preserved for us in the Books of Judges.

When I look out into the night sky, I cannot discern the outline of a lion in the constellation Leo. I can make out the Big Dipper, and, if the night is clear, the Little Dipper. I am at a loss to make out much of a hunter in Orion or a fish

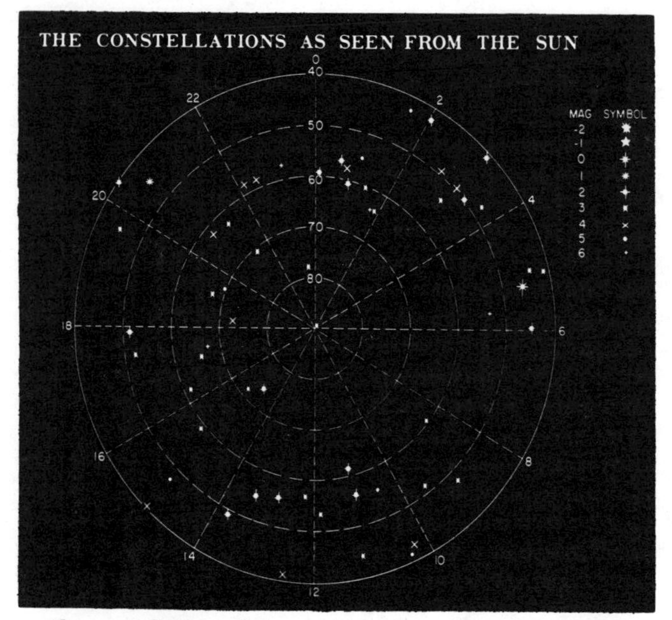

The constellations of the northern sky as seen from the vicinity of the Sun or the Earth.

in the constellation of Pisces, to say nothing of a charioteer in Auriga. The mythical beasts, personages, and instruments placed by men in the sky are arbitrary, not obvious. There are agreements about which constellation is which—sanctioned in recent years by the International Astronomical Union, which draws boundaries separating one constellation from another. But there are few clear pictures in the sky.

These constellations, while drawn in two dimensions, are fundamentally in three dimensions. A constellation, such as

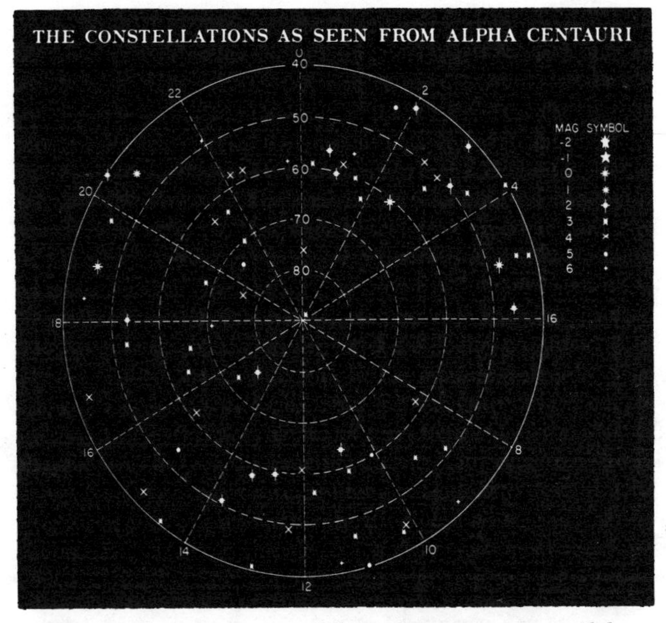

The same scene as viewed from the nearest star, Alpha Centauri. The new star in the constellation of Cassiopeia, near 60 degrees celestial latitude and 2.5 hours celestial longitude, is the Sun.

Orion, is composed of bright stars at considerable distances from Earth and dim stars much closer. Were we to change our perspective, move our point of view—with, for example, an interstellar space vehicle—the appearance of the sky would change. The constellations would slowly distort.

Largely through the efforts of David Wallace at the Laboratory for Planetary Studies at Cornell University, an electronic

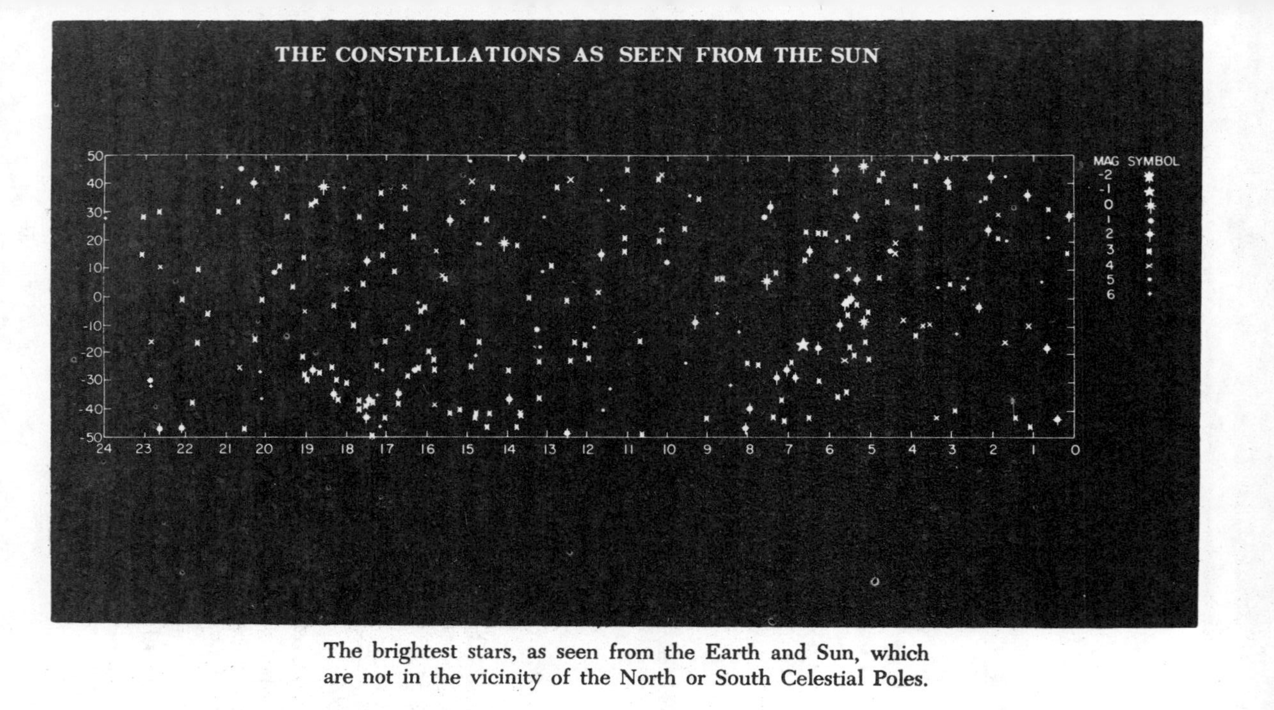

The brightest stars, as seen from the Earth and Sun, which are not in the vicinity of the North or South Celestial Poles.

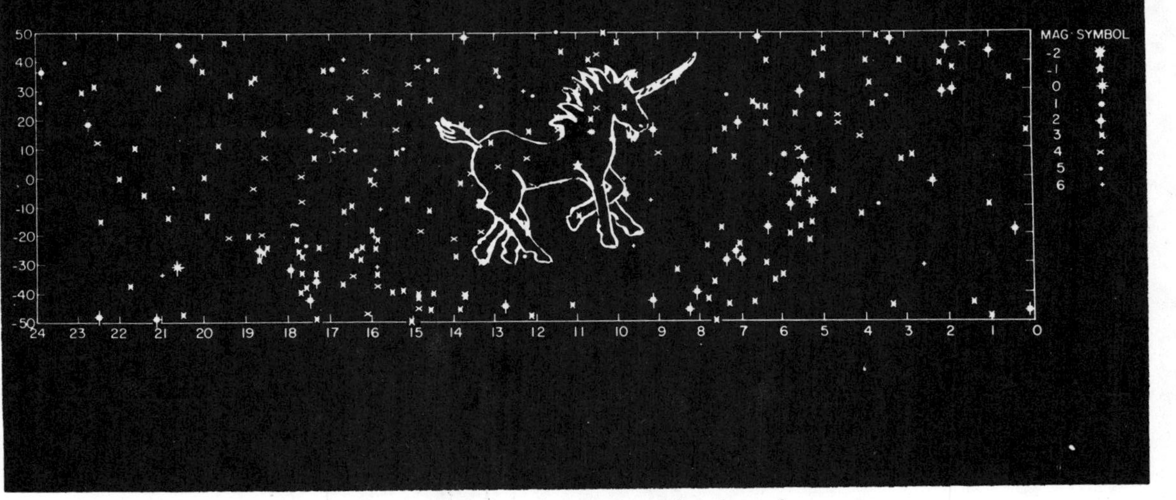

The same stars as seen on opposite page but from the vantage point of Tau Ceti, one of the nearest stars like the Sun. In the sky of Tau Ceti, the Sun is a fourth magnitude star.

computer has been programmed with the information on the three-dimensional positions from the Earth to each of the brightest and nearest stars—down to about fifth magnitude, the limiting brightness visible to the naked eye on a clear night. When we ask the computer to show us the appearance of the sky from Earth, we see results of the sort displayed in the accompanying figures: One for the northern circumpolar constellations, including the Big Dipper, the Little Dipper, and Cassiopeia; one for the southern circumpolar constellations, including the Southern Cross; and one for the broad range of stars at middle celestial latitudes, including Orion and the constellations of the zodiac. If you are not a student of the conventional constellations, you will, I believe, have some difficulty making out scorpions or virgins in the picture.

We now ask the computer to draw us the sky from the nearest star to our own, Alpha Centauri, a triple-star system, about 4.3 light-years from Earth. In terms of the scale of our Milky Way Galaxy, this is such a short distance that our perspectives remain almost exactly the same. From α Cen the Big Dipper appears just as it does from Earth. Almost all the other constellations are similarly unchanged. There is one striking exception, however, and that is the constellation Cassiopeia. Cassiopeia, the queen of an ancient kingdom, mother of Andromeda and mother-in-law of Perseus, is mainly a set of five stars arranged as a W or an M, depending on which way the sky has turned. From Alpha Centauri, however, there is one extra jog in the M; a sixth star appears in Cassiopeia, one significantly brighter than the other five. That star is the Sun. From the vantage point of the nearest star, our Sun is a relatively bright but unprepossessing point in the night sky. There is no way to tell by looking at Cassiopeia from the sky of a hypothetical planet of Alpha Centauri that there are planets going around the Sun, that on the third of these planets there are life forms, and that one of these life forms considers itself to be of quite considerable intelligence. If this is the case for the sixth star in Cassiopeia, might it not also be the case for innumerable millions of other stars in the night sky?

One of the two stars that Project Ozma examined a decade ago for possible extraterrestrial intelligent signals was Tau Ceti, in the constellation (as seen from Earth) of Cetus, the whale. In the accompanying figure, the computer has drawn the sky as seen from a hypothetical planet of τ Cet. We are now a little more than eleven light-years away from the Sun. The perspective has changed somewhat more. The relative orientation of the stars has varied, and we are free to invent new constellations—a psychological projective test for the Cetians.

I asked my wife, Linda, who is an artist, to draw a constellation of a unicorn in the Cetian sky. There is already a unicorn in our sky, called Monoceros, but I wanted this to be a larger and more elegant unicorn—and also one slightly different from common terrestrial unicorns—with six legs, say, rather than four. She invented quite a handsome beast. Contrary to my expectation that he would have three pairs of legs, he is quite proudly galloping on two clusters of three legs each, one fore and one aft. It seems quite a believable gallop. There is a tiny star that is just barely seen at the point where the unicorn's tail joins the rest of his body. That faint and uninspiringly positioned star is the Sun. The Cetians may consider it an amusing speculation that a race of intelligent beings lives on a planet circling the star that joins the unicorn to his tail.

When we move to greater distances from the Sun than Tau Ceti—to forty or fifty light-years—the Sun dwindles still further in brightness until it is invisible to an unaided human eye. Long interstellar voyages—if they are ever undertaken—will not use dead-reckoning on the Sun. Our mighty star, on which all life on Earth depends, our Sun, which is so bright that we risk blindness by prolonged direct viewing, cannot be seen at all at a distance of a few dozen light-years—a thousandth of the distance to the center of our Galaxy.

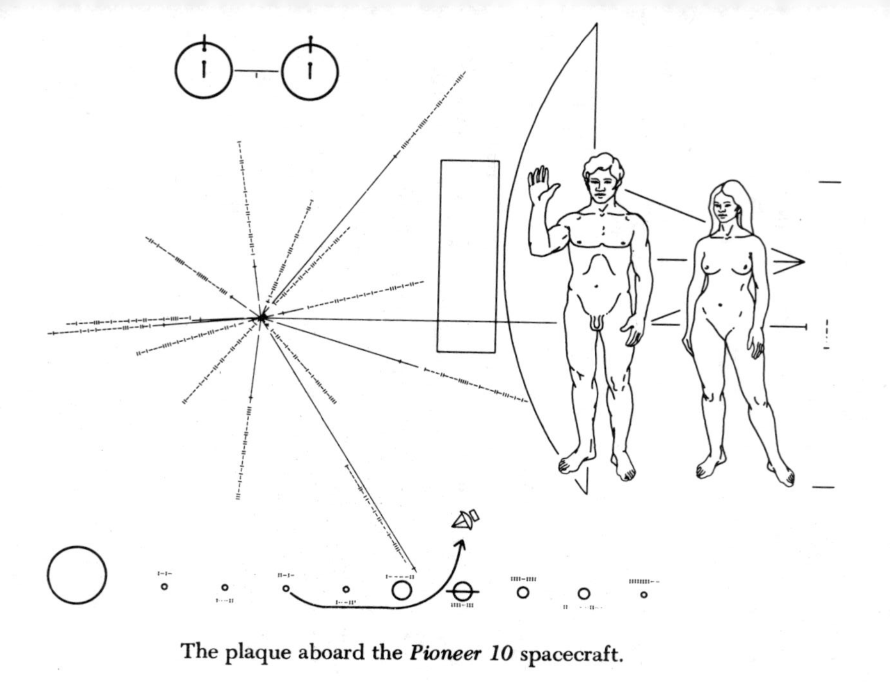

The plaque aboard the *Pioneer 10* spacecraft.

3.
A Message from Earth

Mankind's first serious attempt to communicate with extraterrestrial civilizations occurred on March 3, 1972, with the launching of the *Pioneer 10* spacecraft from Cape Kennedy. *Pioneer 10* was the first space vehicle designed to explore the environment of the planet Jupiter and, earlier in its voyage, the asteroids that lie between the orbits of Mars and Jupiter. Its orbit was not disturbed by an errant asteroid—the safety factor was estimated as 20 to 1. It approached Jupiter on December 3, 1973, and then was accelerated by Jupiter's gravity to become the first man-made object to leave the Solar System. Its exit velocity is about 7 miles per second.

Pioneer 10 is the speediest object launched to date by mankind. But space is very empty, and the distances between the stars are vast. In the next 10 billion years, *Pioneer 10* will not enter the planetary system of any other star, even assuming that all the stars in the Galaxy have such planetary systems. The spacecraft will take about 80,000 years merely to travel the distance to the nearest star, about 4.3 light-years away.

But *Pioneer 10* is not directed to the vicinity of the nearest star. Instead, it will be traveling toward a point on the celestial sphere near the boundary of the constellations Taurus and Orion, where there are no nearby objects.

It is conceivable that the spacecraft will be encountered by an extraterrestrial civilization only if such a civilization has an extensive capability for interstellar space flight and is able to intercept and recover such silent space derelicts.

Placing a message aboard *Pioneer 10* is very much like a shipwrecked sailor casting a bottled message into the ocean—

but the ocean of space is much vaster than any ocean on Earth.

When my attention was drawn to the possibility of placing a message in a space-age bottle, I contacted the *Pioneer 10* project office and NASA headquarters to see if there were any likelihood of implementing this suggestion. To my surprise and delight, the idea met with approval at all steps up the NASA hierarchy, despite the fact that it was—by ordinary standards—very late to make even tiny changes in the spacecraft. During a meeting of the American Astronomical Society in San Juan, Puerto Rico, in December 1971, I discussed privately various possible messages with my colleague Professor Frank Drake, also of Cornell. In a few hours we decided tentatively on the contents of the message. The human figures were added by my artist wife, Linda Salzman Sagan. We do not think it is the optimum conceivable message for such a purpose: There were a total of only three weeks for the presentation of the idea, the design of the message, its approval by NASA, and the engraving of the final plaque. An identical plaque has been launched in 1973 on the *Pioneer 11* spacecraft, on a similar mission.

On the title page of this chapter is shown the message. It is etched on a 6-inch by 9-inch gold-anodized aluminum plate, attached to the antenna support struts of *Pioneer 10*. The expected erosion rate in interstellar space is sufficiently small that this message should remain intact for hundreds of millions of years, and probably for a much longer period of time. It is, thus, the artifact of mankind with the longest expected lifetime.

The message itself intends to communicate the locale, epoch, and something of the nature of the builders of the spacecraft. It is written in the only language we share with the recipients: Science. At top left is a schematic representation of the hyperfine transition between parallel and antiparallel proton and electron spins of the neutral hydrogen atom. Beneath this representation is the binary number 1. Such transitions of hydrogen are accompanied by the emission of a radio-frequency photon of wavelength about 21 centimeters and frequency of about 1,420 Megahertz. Thus, there is a characteristic distance and a characteristic time associated with the transition. Since hydrogen is the most abundant atom in the Galaxy, and physics is the same throughout the Galaxy, we think there will be no difficulty for an advanced civilization to understand this part of the message. But as a check, on the right margin is the binary number 8 (1---) between two tote marks, indicating the height of the *Pioneer 10* spacecraft, schematically represented behind the man and the woman. A civilization that acquires the plaque will, of course, also acquire the spacecraft, and will be able to determine that the distance indicated is indeed close to 8 times 21 centimeters, thus confirming that the symbol at top left represents the hydrogen hyperfine transition.

Further binary numbers are shown in the radial pattern comprising the main part of the diagram at left center. These numbers, if written in decimal notation, would be ten digits long. They must represent either distances or times. If distances, they are of the order of several times 10^{11} centimeters, or a few dozen times the distance between the Earth and the Moon. It is highly unlikely that we would consider them useful to communicate. Because of the motion of objects within the Solar System, such distances vary in continuous and complex ways.

However, the corresponding times are on the order of 1/10 second to 1 second. These are the characteristic periods of the pulsars, natural and regular sources of cosmic radio emission; pulsars are rapidly rotating neutron stars produced in catastrophic stellar explosions (see Chapter 38). We believe that a scientifically sophisticated civilization will have no difficulty understanding the radial burst pattern as the positions and periods of 14 pulsars with respect to the Solar System of launch.

But pulsars are cosmic clocks that are running down at largely known rates. The recipients of the message must ask themselves not only where it was ever possible to see 14 pulsars arrayed in such a relative position, but also *when* it was possible to see them. The answers are: Only from a very small volume of the Milky Way Galaxy and in a single year in the history of the Galaxy. Within that small volume there are perhaps a thousand stars; only one is anticipated to have the array of planets with relative distances as indicated at the bottom of the diagram. The rough sizes of the planets and the rings of Saturn are also schematically shown. A schematic representation of the initial trajectory of the spacecraft launched from Earth and passing by Jupiter is also displayed. Thus, the message specifies one star in about 250 billion and one year (1970) in about 10 billion.

The content of the message to this point should be clear to an advanced extraterrestrial civilization, which will, of course, have the entire *Pioneer 10* spacecraft to examine as well. The message is probably less clear to the man on the street, if the street is on the planet Earth. (However, scientific communities on Earth have had little difficulty decoding the message.) The opposite is the case with the representations of human beings to the right. Extraterrestrial beings, which are the product of 4.5 billion years or more of independent biological evolution, may not at all resemble humans, nor may the perspective and line-drawing conventions be the same there as here. The human beings are the most mysterious part of the message.

4.
A Message to Earth

The golden greeting card placed aboard the *Pioneer 10* spacecraft was intended for the remote contingency that representatives of an advanced extraterrestrial civilization, some time in the distant future, might encounter this first artifact of mankind to leave the Solar System. But the message has had a more immediate impact. It has already been meticulously studied—not by extraterrestrials, but by terrestrials. Human beings all over the planet Earth have examined the message, applauded it, criticized it, interpreted it, and proposed alternative messages.

The graphics of the message have been reproduced widely in newspapers and television programs, small art and literary magazines, and national newsweeklies. We have received letters from scientists and housewives, historians and artists, feminists and homosexuals, military and foreign service officers, and one professor of bass fiddle. Our plaque has been reproduced for commercial sale by an engraving company, a distributor of scientific knickknacks, a manufacturer of tapestry, and an Italian mint specializing in silver ingots—all, incidentally, without authorization.

The great majority of comments have been favorable, some extraordinarily enthusiastic. The large street advertising billboards for the *Tribune* of Geneva, Switzerland, announced "*Message de la NASA pour les extraterrestres!*" One scientist writes to say that the description of the scientific basis of the plaque we published in the American journal *Science* was the first scientific paper he had ever read that moved him to tears of joy. A correspondent in Athens, Georgia, writes, "We'll all be gone before this particular message in a bottle is picked up by some indescribable spacecomber; nevertheless, its very existence, the audacity of the dream, inevitably produces in me—and many others I know—the feelings of a Balboa, a Leeuwenhoek, a human being being human!"

At the California Institute of Technology, where the graffiti is arcane, some unknown artist drew the message life-size on a barrier at a building site, eliciting friendly greetings from the inhabitants, which we hope will serve as a model for extraterrestrial readers (see the illustration on the facing page).

But there were also critical comments. They were not directed at the pulsar map, which was the scientific heart of the message, but rather at the representation of the man and the woman. The original drawings of this couple were made by my wife and were based upon the classical models of Greek sculpture and the drawings of Leonardo da Vinci. We do not think this man and woman are ignoring each other. They are not shown holding hands lest the extraterrestrial recipients believe that the couple is one organism joined at the fingertips. (In the absence of indigenous horses, both the Aztecs and the Incas interpreted the mounted *conquistador* as one animal—a kind of two-headed centaur.) The man and woman are not shown in precisely the same position or carriage so that the suppleness of the limbs could be communicated—although we well understand that the conventions of perspective and line drawing popular on Earth may not be readily apparent to civilizations with other artistic conventions.

The man's right hand is raised in what I once read in an anthropology book is a "universal" sign of good will—although any literal universality is of course unlikely. At least the greeting displays our opposable thumbs. Only one of the two people is shown with hand raised in greeting, lest the recipients deduce erroneously that one of our arms is bent permanently at the elbow.

Several women correspondents complain that the woman appears too passive. One writes that she also wishes to greet

Graffiti at Caltech: A response to the *Pioneer 10* plaque. Courtesy "Engineering and Science," California Institute of Technology, Pasadena, Calif.

the universe, with both arms outstretched in womanly salutation. The principal feminine criticism is that the woman is drawn incomplete—that is, without any hint of external genitalia. The decision to omit a very short line in this diagram was made partly because conventional representation in Greek statuary omits it. But there was another reason: Our desire to see the message successfully launched on *Pioneer 10*. In retrospect, we may have judged NASA's scientific-political hierarchy as more puritanical than it is. In the many discussions that I held with such officials, up to the Administrator of the National Aeronautics and Space Administration and the President's Science Adviser, not one Victorian demurrer was ever voiced; and a great deal of helpful encouragement was given.

Yet it is clear that at least some individuals were offended even by the existing representation. The Chicago *Sun Times*, for example, published three versions of the plaque in different editions all on the same day: In the first the man was represented whole; in the second, suffering from an awkward and botched airbrush castration; and in the final version—intended no doubt to reassure the family man dashing home—with no sexual apparatus at all. This may have pleased one feminist correspondent who wrote to the New York *Times* that she was so enraged at the incomplete representation of the woman that she had an irresistible urge "to cut off the man's . . . right arm!"

The Philadelphia *Inquirer* published on its front page an illustration of the plaque, but with the nipples of the woman and the genitalia of the man removed. The assistant managing editor was quoted as saying, "A family newspaper must uphold community standards."

An entire mythology has evolved about the absence of discernible female genitalia. It was a column by the respected science writer Tom O'Toole, of the Washington *Post*, that first reported that NASA officials had censored an original depiction of the woman. This tale was then circulated in nationally syndicated columns by Art Hoppe, Jack Stapleton, Jr., and others. Stapleton imagined the enraged citizens of another planet receiving the plaque, and in a paroxysm of moral outrage covering over with adhesive tape the pornographic representation of the *feet* of the man and the woman. One letter writer to the Washington *Daily News* proposed that if the woman was to be censored, then for consistency the noses of the humans should have been painted blue. A tut-tutting letter in *Playboy* magazine complained about this further intrusion of government censorship, already quite bad enough, into the lives of the citizenry. Editorials in science-fiction magazines also took the government to task. The idea of government censorship of the *Pioneer 10* plaque is now so well documented and firmly entrenched that no statement from the designers of the plaque to the contrary can play any role in influencing the prevailing opinion. But we can at least try.

What sexuality there is in the message also drew epistolary fire. The Los Angeles *Times* published a letter from an irate reader that went:

> I must say I was shocked by the blatant display of both male and female sex organs on the front page of the *Times*. Surely this type of sexual exploitation is below the standards our community has come to expect from the *Times*.
>
> Isn't it enough that we must tolerate the bombardment of pornography through the media of film and smut magazines? Isn't it bad enough that our own space agency officials have found it necessary to spread this filth even beyond our own solar system?

This was followed several days later by another letter in the *Times:*

> I certainly agree with those people who are protesting our sending those dirty pictures of naked people out into space. I think the way it should have been done would have been to visually bleep out the reproductive organs

of the drawings of the man and the woman. Next to them should have been a picture of a stork carrying a little bundle from heaven.

Then if we really want our celestial neighbors to know how far we have progressed intellectually, we should have included pictures of Santa Claus, the Easter Bunny, and the Tooth Fairy.

The New York *Daily News* headlined the story in typical fashion: "Nudes and Map tell about Earth to Other Worlds."

Some correspondents argue that the function of the sexual organs would not be obvious even had they been graphically displayed, and urged on us a sequence of cartoons from copulation to birth to puberty to copulation. There was not quite room for this on a 6-inch by 9-inch plaque. I can also imagine the letters that would then have been written to the Los Angeles *Times*.

An article in *Catholic Review* criticizes the plaque because it "includes everything but God," and suggests that, rather than a pair of human beings, it would have been better to have borne a sketch of a pair of praying hands.

Another correspondent maintains that the perspective conventions are insuperably difficult, and urges us to send the complete cadavers of a man and a woman. They would be perfectly preserved in the cold of space, and could be examined by extraterrestrials in detail. We declined on grounds of excess weight.

The front page of the Berkeley, California, *Barb*, apparently intending to convey an impression that the man and woman on the message were too straight, reproduced them with the caption, "Hello. We're from Orange County."

This comment touches on an aspect of the representation of the man and woman that I personally feel much worse about, although it has received almost no other public notice. In the original sketches from which the engravings were made, we made a conscious attempt to have the man and woman panracial. The woman was given epicanthian folds and in other ways a partially Asian appearance. The man was given a broad nose, thick lips, and a short "Afro" haircut. Caucasian features were also present in both. We had hoped to represent at least three of the major races of mankind. The epicanthian folds, the lips, and the nose have survived into the final engraving. But because the woman's hair is drawn only in outline, it appears to many viewers as blond, thereby destroying the possibility of a significant contribution from an Asian gene pool. Also, somewhere in the transcription from the original sketch drawing to the final engraving the Afro was transmuted into a very non-African Mediterranean-curly haircut. Nevertheless, the man and woman on the plaque are, to a significant degree, representative of the sexes and races of mankind.

Professor E. Gombrich, the Director of the Warburg Institute, a leading art school in London, criticizes the plaque in the journal *Scientific American.* He wonders how the plaque can possibly be expected to be visible to an extraterrestrial organism that may not have developed the sense of sight at visible wavelengths. The answer is derived simply from the laws of physics. Planetary atmospheres absorb light from the nearby sun or suns because of three molecular processes. The first is a change in the energy state of individual electrons attached to atoms. These transitions occur in the ultraviolet, X-ray, and gamma-ray parts of the spectrum and tend to make a planetary atmosphere opaque at these wavelengths. Second, there are vibrational transitions that occur when two atoms in a given molecule oscillate with respect to each other. Such transitions tend to make planetary atmospheres opaque in the near infrared part of the spectrum. Third, molecules undergo rotational transitions, due to the free rotation of the molecule. Such transitions tend to absorb in the far infrared. As a result, quite generally, the radiation from the nearby star, which penetrates through a planetary atmosphere, will be in the visible and in the radio parts of the spectrum—the parts that are not absorbed by the atmosphere.

In fact, these are the principal "windows" that astronomers use for surveying the universe from the Earth's surface. But radio wavelengths are so long that no organisms of reasonable size can develop pictures of their surroundings with radio wavelength "eyes." Therefore, we expect optical frequency sensors to be developed quite widely among organisms on planets of stars throughout the Galaxy.

However, even if we imagine organisms whose eyes work in the infrared region (or, for that matter, in the gamma-ray region) and who are able to intercept *Pioneer 10* in interstellar space, it is probably not asking too much of them to have contrivances that scan the plaque at frequencies to which their eyes are insensitive. Because the engraved lines on the plaque are darker than the surrounding gold-anodized aluminum, the message should be entirely visible even in the infrared.

Gombritch also takes us to task for portraying an arrow as a sign of the spacecraft's trajectory. He maintains that arrows would be understandable only to civilizations that have evolved, as ours has, from a hunting society. But here again it does not take a very intelligent extraterrestrial to understand the meaning of the arrow. There is a line that begins on the third planet of a solar system and ends, somewhere in interstellar space, at a schematic representation of the spacecraft —which the discoverers of the message have at "hand": The plaque is attached to the spacecraft. From this I would hope they would be able to argue backward to our hunter-gatherer ancestors.

In the same way, the relative distances of the planets from the Sun, shown by binary notation at the bottom of the plaque, indicate that we use base-10 arithmetic. From the fact that we have 10 fingers and 10 toes—drawn with some care on the plaque—I hope any extraterrestrial recipients will be able to deduce that we use base-10 arithmetic and that some of us count on our fingers. From the stumpiness of our toes they may even be able to deduce that we evolved from arboreal ancestors.

There are other respects in which the message has proved to be a psychological projective test. One man writes of his concern that the message has doomed all of mankind. American movies of Second World War vintage, he argues, are very likely propagating via television transmission through interstellar space. From such programming, the extraterrestrials will easily be able to deduce (1) that the Nazis were very bad fellows, and (2) that they greeted each other with their right hand extended outward. From the fact that the man on the plaque is portrayed as making what our correspondent erroneously perceives as the same sort of greeting, he is concerned that the extraterrestrials will deduce that the wrong side won World War II and promptly mount a punitive expedition to Earth to set matters straight.

Such a letter more nearly describes the state of mind of the writer than of the likely extraterrestrial recipients of the message. The raised right hand in greeting is historically connected with militarism, but in a negative way: The raised and empty right hand symbolizes that no weapon is being carried.

For me, some of the most moving responses to the message are the works of art and poetry that it evoked. Mr. 'Aim Morhardt is a painter of water colors of the desert and sierras who lives in Bishop, California, where, perhaps not coincidentally, the giant Goldstone tracking station, which commands *Pioneer 10*, is located. Mr. Morhardt's poem follows:

Pioneer 10: The Golden Messenger.

The dragon prows that cruised the northern seas,
Questing adventure with the fighting clan;
The gallant mermaid bows blown down the breeze
On barquentine and slim-hulled merchantman;
All the discoverers of unknown lands
Gone in this winged age where naught remains
Of new strange treasure on some foreign strand,
So well-known earth, such charted routes and lanes.

Now the new figurehead of man appears,
Facing the vast immeasurable unknown,
Naked, star-sped, beyond the call of years,
Hand in hand, outward bound, and so alone.
Go, tiny messenger of our your race,
Touch, if you can, harbor in some far place.

Mr. Arvid F. Sponberg, of Belfast, Northern Ireland, writes: "The voyage of *Pioneer 10*—and the voyages of those like her—will have an effect that poets, painters and musicians will not long ignore. The existence of the *idea* of *Pioneer 10* is proof of this. The scientific mission of course is of incalculable value and interest, but the *idea* of the journey is of even greater imaginative value. *Pioneer 10* brings closer the day when artists must confront man's new voyage as experience and not fantasy."

Mr. Sponberg composed for us a poem in sonnet form:

New Odyssey

Away, afar, beyond, bereft of kin,
Wayward, wandering, far ranging vagabonds,
Yearning, stardrawn, the Pioneers sweep on,
Outward bound, adrift on the solar wind.

A man, a woman, orphans of warm earth
Or splendid voyageurs with golden sails,
Or gypsies roaming ancient stellar trails,
A caravan in quest of celestial berth.

If, deep within cold interstellar space,
Some fearful eye spies life on this raft,
Will it perceive the heart within our craft,
A pulsar pounding out the rhythms of peace?

A spirit's starburst pierces new frontiers;
An Odyssey is our home; let us praise Pioneers!

There is, of course, the possibility that the message on *Pioneer 10*—invented by human beings but directed at creatures of a very different kind—may prove ultimately mysterious to them. We think not. We think we have written the message—except for the man and woman—in a universal language. The extraterrestrials cannot possibly understand English or Russian or Chinese or Esperanto, but they must share with us common mathematics and physics and astronomy. I believe that they will understand, with no very great effort, this message written in the galactic language: "Scientific."

But we may be wrong. One exploration of a total misunderstanding—and by far the most amusing such description—was made by the British humor magazine *Punch* in an article headlined, "According to the [Paris] *Herald Tribune* only one in ten of NASA scientists was able to figure out its message. So what chance have the aliens got?" *Punch* presents an opinion sampling of four representative extraterrestrials. They should be read with close reference to the illustration of the actual message:

"Still, I must emphasise that we are only guessing at this stage and none of us has been able to explain the significance of the dots along the bottom. A suggestion that it could be a map of some metropolitan railway has been made to us, but we feel that this fails to take into account the arrowed position of a capsized yacht, or possibly a garden trowel. The inclusion of a naked blonde makes it more than likely, however, that this is some kind of a joke sent out by a backward planet, possibly that being used by the Earthlings."

"Speaking as a fourteen-legged and extremely thin spider," said a voice from the back of Andromeda 9, "I have studied this post-card from the Earthlings and I take it as a snub. The caricature of our species is both crude and inept, suggesting, amongst other things, that we've got a right leg longer than all the rest. Furthermore, the geometric being which is standing at the back has clearly turned its back on us and one of the other two is pointing five antennae in a frankly sordid gesture. There seems little reason to doubt, amongst us intelligent spiders, that this thing is intended as a declaration of war. The illustrated talent for the creature on the right to be capable of firing arrows from the shoulder is a particularly sinister turn and one that bodes badly for a long and bitter struggle with the Earthlings."

"Whatever it is," the Being declared, "it's not come all this way for nothing. My guess is that it's trying to tell us something. Just suppose, for argument's sake, that this thing which we have before us is not an actual creature itself but an artifact of some sort. Such a theory might explain for a start why it hasn't so far uttered in any way. No, this thing was sent—probably from some primitive three-dimensional world—and I say it's meant to be a picture or a cipher with a message for us Beings. What the message is, of course, depends on which way up it's supposed to be. I shouldn't be a bit surprised if it was rude."

"Magnificent!" The thing on Alpha Centaurus was overcome with awe. "Truly magnificent! As far as is known, this is the first time ever that has fetched up on our planet an original work by the erstwhile Earthling Leonardo da Vinci! Our telescopes show that the style is unmistakably his. Nevertheless, the discovery is bound to alter some of intelligence data on the Earth. It was not known, until now, that the climate was sufficiently warm for policemen to go to point duty without clothes in their world nor that key limbs on the Earthlings are apparently operated by string. Let us hope they send us further simple greeting-cards soon."

Perhaps the most perceptive editorial comment is the New York *Times*':

> . . . that gold-plated plaque is more of a challenge to us. Despite the uncanny mastery of celestial laws that permits man to shoot his artifacts at the stars, we find ourselves still depressingly inept at ordering our own systems here on Earth. Even as we try to find a way to insure that sapient man will not consume his planet in nuclear fires, a rising chorus warns us that man may very well exhaust his earth either by overbreeding or by inordinate demands on its resources, or both.
>
> So the marker launched into space is at the same time a gauntlet thrown down to earth: That the gold-plated plaque convey in its time the message that man is still here—not that he had been here.

The message aboard *Pioneer 10* has been good fun. But it has been more than that. It is a kind of cosmic Rorschach test, in which many people see reflected their hopes and fears, their aspirations and defeats—the darkest and the most luminous aspects of the human spirit.

The sending of such a message forces us to consider how we wish to be represented in a cosmic discourse. What is the image of mankind that we might wish to represent to a superior civilization elsewhere in the Galaxy? The transmittal of the *Pioneer 10* message encourages us to consider ourselves in cosmic perspective.

The greater significance of the *Pioneer 10* plaque is not as a message to out there; it is as a message to back here.

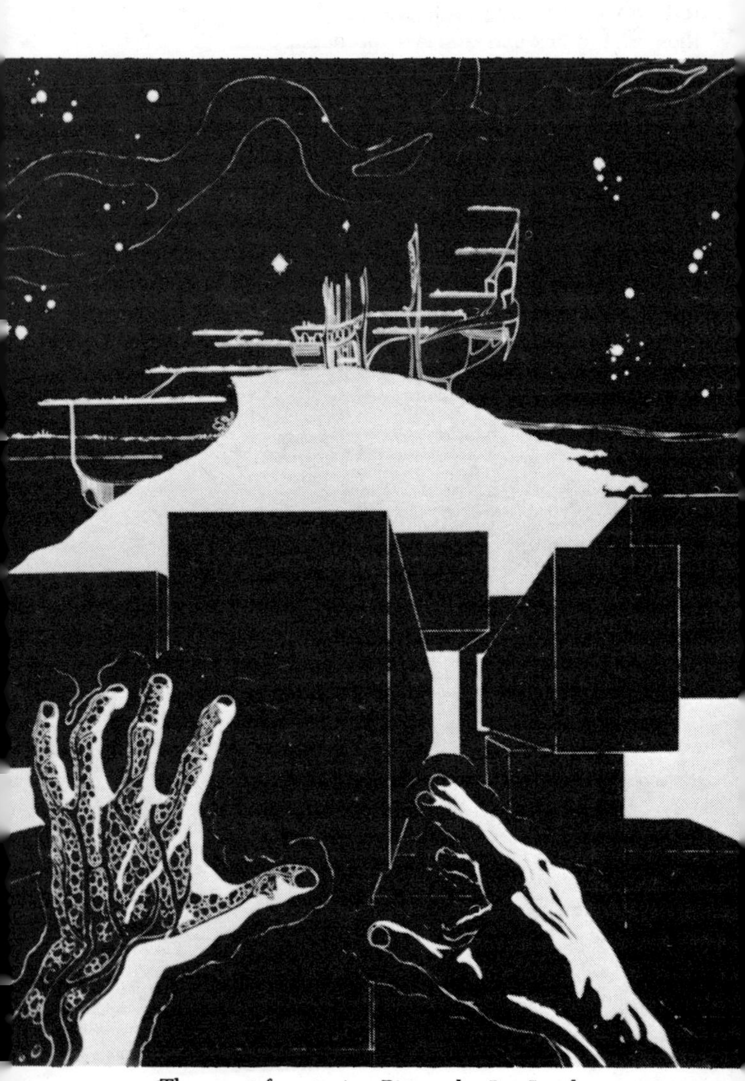

The quest for utopias. Picture by Jon Lomberg.

5. Experiments in Utopias

In assessing the likelihood of advanced technical civilizations elsewhere in the Galaxy, the most important fact is the one about which we know least—the lifetime of such a civilization. If civilizations destroy themselves rapidly after reaching the technological phase, at any given moment (like now) there may be very few of them for us to contact. If, on the other hand, a small fraction of civilizations learn to live with weapons of mass destruction and avoid both natural and self-generated catastrophes, the number of civilizations for us to communicate with at any given moment may be very large.

This assessment is one reason we are concerned about the lifetime of such civilizations. There is a more pressing reason, of course. For personal reasons, we hope that the lifetime of our own civilization will be long.

There is probably no epoch in the history of mankind that has undergone so much and so many varieties of change as the present time. Two hundred years ago, information could be sent from one city to another no faster than by horse. Today, the information can be sent via telephone, telegraph, radio, or television at the velocity of light. In two hundred years the speed of communication has increased by a factor of thirty million. We believe there will be no corresponding future advance, since messages cannot, we believe, be sent faster than the velocity of light.

Two hundred years ago it took as long to go from Liverpool to London as it now does from the Earth to the Moon. Similar changes have occurred in the energy resources available to our civilization, in the amount of information that is stored and processed, in methods of food production and distribution, in the synthesis of new materials, in the concentration of population from the countryside to the cities, in the vast increase in population, in improved medical practice, and in enormous social upheaval.

Our instincts and emotions are those of our hunter-gatherer ancestors of a million years ago. But our society is astonishingly different from that of a million years ago. In times of slow change, the insights and skills learned by one generation are useful, tried, and adaptive, and are gladly received when passed down to the next generation. But in times like today, when the society changes significantly in less than a human lifetime, the parental insights no longer have unquestioned validity for the young. The so-called generation gap is a consequence of the rate of social and technological change.

Even within a human lifetime, the change is so great that many people are alienated from their own society. Margaret Mead has described older people today as involuntary immigrants from the past to the present.

Old economic assumptions, old methods of determining political leaders, old methods of distributing resources, old methods of communicating information from the government to the people—and vice versa—all of these may once have been valid or useful or at least somewhat adaptive, but today may no longer have survival value at all. Old oppressive and chauvinistic attitudes among the races, between the sexes, and between economic groups are being justifiably challenged. The fabric of society throughout the world is ripping.

At the same time, there are vested interests opposed to change. These include individuals in power who have much to gain in the short run by maintaining the old ways, even if their children have much to lose in the long run. They are individuals who are unable in middle years to change the attitudes inculcated in their youth.

The situation is a very difficult one. The rate of change cannot continue indefinitely; as the example of the rate of communication indicates, limits must be reached. We cannot communicate faster than the velocity of light. We cannot have a population larger than Earth's resources and economic distribution facilities can maintain. Whatever the solutions to be achieved, hundreds of years from now the Earth is unlikely still to be experiencing great social stress and change. We will have reached some solution to our present problems. The question is, which solution?

In science a situation as complicated as this is difficult to treat theoretically. We do not understand all the factors that influence our society and, therefore, cannot make reliable predictions on what changes are desirable. There are too many complex interactions. Ecology has been called the subversive science because every time a serious effort to preserve a feature of the environment is made, it runs into enormous numbers of social or economic vested interests. The same is true every time we attempt to make a major change in anything that is wrong; the change runs through society as a whole. It is difficult to isolate small fragments of the society and change them without having profound influences on the rest of society.

When theory is not adequate in science, the only realistic approach is experimental. Experiment is the touchstone of science on which the theories are framed. It is the court of last resort. What is clearly needed are experimental societies!

There is good biological precedent for this idea. In the evolution of life there are innumerable cases when an organism was clearly dominant, highly specialized, perfectly acclimatized to its environment. But the environment changed and the organism died. It is for this reason that nature employs mutations. The vast majority of mutations are deleterious or lethal. The mutated species are less adaptive than the normal types. But one in a thousand or one in ten thousand mutants has a slight advantage over its parents. The mutations breed true, and the mutant organism is now slightly better adapted.

Social mutations, it seems to me, are what we need. Perhaps because of a hoary science-fiction tradition that mutants are ugly and hateful, it might be better to use another term. But social mutation—a variation on a social system which breeds true, which, if it works, is the path to the future—seems to be precisely the right phrase. It would be useful to examine why some of us find the phrase objectionable.

We should be encouraging social, economic, and political experimentation on a massive scale in all countries. Instead, the opposite seems to be occurring. In countries such as the United States and the Soviet Union the official policy is to discourage significant experimentation, because it is, of course, unpopular with the majority. The practical consequence is vigorous popular disapproval of significant variation. Young urban idealists immersed in a drug culture, with dress styles considered bizarre by conventional standards, and with no prior knowledge of agriculture, are unlikely to succeed in establishing utopian agricultural communities in the American Southwest—even without local harassment. Yet such experimental communities throughout the world have been subjected to hostility and violence by their more conventional neighbors. In some cases the vigilantes are enraged because they themselves have only within the previous generation been accepted into the conventional system.

We should not be surprised, then, if experimental communities fail. Only a small fraction of mutations succeed. But the advantage social mutations have over biological mutations is that individuals learn; the participants in unsuccessful communal experiments are able to assess the reasons for failure and can participate in later experiments that attempt to avoid the causes of initial failure.

There should be not only popular approval for such experiments, but also official governmental support for them. Volunteers for such experiments in utopia—facing long odds for the benefit of society as a whole—will, I hope, be thought of as men and women of exemplary courage. They are the cutting edge of the future. One day there will arise an experimental community that works much more efficiently than the polyglot, rubbery, hand-patched society we are living in. A viable alternative will then be before us.

I do not believe that anyone alive today is wise enough to know what such a future society will be like. There may be many different alternatives, each potentially more successful than the pitifully small variety that face us today.

A related problem is that the non-Western, nontechnological societies, viewing the power and great material wealth of the West, are making great strides to emulate us—in the course of which many ancient traditions, world-views, and ways of life are being abandoned. For all we know, some of the alternatives being abandoned contain elements of precisely the alternatives we are seeking. There must be some way to preserve the adaptive elements of our societies—painfully worked out through thousands of years of sociological evolution—while at the same time coming to grips with modern technology. The principal immediate problem is to spread the technological achievements while maintaining cultural diversity.

An opinion sometimes encountered is that the problem is technology itself. I maintain that it is the misuse of technology by the elected or self-appointed leaders of societies, and not technology itself, that is at fault. Were we to return to more primitive agricultural endeavors, as some have urged, and abandon modern agricultural technology, we would be condemning hundreds of millions of people to death. There is no escape from technology on our planet. The problem is to use it wisely.

For quite similar reasons, technology must be a major factor in planetary societies older than ours. I think it likely that societies that are immensely wiser and more benign than ours are, nevertheless, more highly technological than we.

We are at an epochal, transitional moment in the history of life on Earth. There is no other time as risky, but no other time as promising for the future of life on our planet.

A comment on chauvinism. Courtesy, Paul Conrad, Los Angeles *Times*.

6. Chauvinism

Jokes are a way of dealing with anxiety. There is a class of jokes dealing with extraterrestrial life. In one, the extraterrestrial visitor lands on Earth, walks up to a gasoline pump or a gumball machine—the accounts differ—and asks, "What's a nice girl like you doing in a place like this?"

Elsewhere, beings are doubtless very different from us. But the joke assumes that extraterrestrial organisms will be, if not like human beings, then like gasoline pumps or gumball machines. The most likely circumstance is that extraterrestrial beings will look nothing like any organisms or machines familiar to us. Extraterrestrials will be the product of billions of years of independent biological evolution, by small steps, each involving a series of tiny mutational accidents, on planets with very different environments from those that characterize Earth.

But such jokes underscore a general problem and a general virtue in thinking about life elsewhere. The problem is that we have only one kind of life to study, the co-related biology of the planet Earth, all organisms of which have descended from a single instance of the origin of life. It is difficult for the biologist, as well as the layman, to determine what properties of life on our planet are accidents of the evolutionary process and what properties are characteristic of life everywhere. The assumption that life elsewhere has to be, in some major sense, like life here is a conceit I will call chauvinism.

While such chauvinism has been common throughout human history, clearer views have occasionally surfaced, for example, by the great French astronomer Pierre Simon, the Marquis de Laplace. In his classic work *La Mecanique Celeste* he wrote: "[The Sun's] influence gives birth to the animals and plants which cover the surface of the Earth, and analogy induces us to believe that it produces similar effects on the planets; for it is not natural to pose that matter, of which we see the fecundity develop itself in such various ways, should be sterile upon a planet so large as Jupiter, which, like the Earth, has its days, its nights, and its years, and on which observation discovers changes that indicate very active forces. Man, formed for the temperature which he enjoys upon the Earth, could not, according to all appearance, live upon the other planets; but ought there not to be a diversity of organization suited to the various temperatures of the globes of this universe? If the difference of elements and climates alone causes such variety in the production of the Earth, how infinitely diversified must be the production of the planets and their satellites?" Laplace wrote these words near the end of the eighteenth century.

The virtue of thinking about life elsewhere is that it forces us to stretch our imaginations. Can we think of alternative solutions to biological problems already solved in one particular way on Earth? For example, the wheel is a comparatively recent invention on the planet Earth. It seems to have been invented in the ancient Near East less than ten thousand years ago. In fact, the high civilizations of Meso-America, the Aztecs and the Mayas, never employed the wheel, except for children's toys. Biology—the evolutionary process—has never invented the wheel, in spite of the fact that its selective advantages are manifest. Why are there no wheeled spiders or goats or elephants rolling along the highways? The answer is clearly that, until recently, there were no highways. Wheels are of use only when there are surfaces to roll on. Since the planet Earth is a heterogeneous, bumpy place with few long, smooth areas, there was no advantage to evolving the wheel. We can very well imagine another planet with enormous long stretches of smooth lava fields in which wheeled organisms are abundant. The late Dutch artist M. C. Escher designed a salamander-like organism that would do very well in such an environment.

The evolution of life on Earth is a product of random events, chance mutations, and individually unlikely steps; small differences early in the evolution of life have a profound significance later in the evolution of life. Were we to start the Earth over again and let only random factors operate, I believe that we would wind up with nothing at all resembling human beings. This being the case, how much less likely it is that organisms evolving over five billion or more years, independently in a quite different environment of another planet of a far-off star, would closely resemble human beings.

Thus, the hoary science-fiction standby of the sexual love between a human being and an inhabitant of another planet ignores, in the most fundamental sense, the biological realities. John Carter could love Dejah Thoris, but, despite what Edgar Rice Burroughs believed, their love could not be consummated. And if it could, a viable offspring would not be possible. Likewise, the category of contact story, now quite fashionable in some UFO enthusiast circles, of sexual contact between human and saucerian—most recently described in a weekly newspaper headline with the modest title "We Sexed a Blonde from a Flying Saucer!"—must be relegated to the realm of improbable fantasy. Such crossings are about as reasonable as the mating of a man and a petunia.

A popular phrase—often encountered in popular books on the planets—is "life as we know it." We read that "life as we know it" is impossible on this planet or that. But what is life as we know it? It depends entirely on who the "*we*" is. A person who is unsophisticated in biology, who lacks a keen appreciation of the multitudinous adaptations and varieties of terrestrial organisms, will have a meager idea of the range of possible biological habitats. There are discussions, even by famous scientists, that give the impression that an environment that is uncomfortable for my grandmother is impossible for life.

At one time it was thought that oxides of nitrogen had been detected in the atmosphere of Mars. A scientific paper was published on this apparent finding. The authors of the paper argued that life on Mars was, therefore, impossible, because oxides of nitrogen are poisonous gases. There are at least two objections to this argument. First, oxides of nitrogen are poisonous gases only to some organisms on Earth. Second, what quantity of oxides of nitrogen were thought to be discovered on Mars? When I calculated the amount, it turned out to be less than the average abundance above Los Angeles. The oxides of nitrogen are an important constituent of smog. Life in Los Angeles may be difficult, but it is not yet impossible. The same conclusion applies to Mars. The final problem with these particular observations is that they are very likely mistaken; later studies—for example, observations Tobias Owen and I made with the Orbiting Astronomical Observatory—have shown no oxides of nitrogen in the atmosphere of Mars.

Oxygen chauvinism is common. If a planet has no oxygen, it is alleged to be uninhabitable. This view ignores the fact that life arose on Earth in the absence of oxygen. In fact, oxygen chauvinism, if accepted, logically demonstrates that life anywhere is impossible. Fundamentally, oxygen is a poisonous gas. It chemically combines with and destroys the organic molecules of which terrestrial life is composed. There are many organisms on Earth that do without oxygen and many organisms that are poisoned by it.

All of the earliest organisms on Earth did not use molecular oxygen, O_2. In a brilliant set of evolutionary adaptations, organisms like insects and frogs and fish and people learned not only to survive in the presence of this poisonous gas but actually to use it to increase the efficiency with which we metabolize foodstuffs. But that should not blind us to the fundamentally poisonous character of this gas. The absence of oxygen on a place such as Jupiter is, therefore, hardly an argument against life on such planets.

There are ultraviolet light chauvinists. Because of the oxygen in the Earth's atmosphere, a variety of oxygen molecule called ozone (O_3) is produced high in the atmosphere, about twenty-five miles above the surface. This ozone layer absorbs the middle-wavelength ultraviolet rays from the Sun, preventing them from reaching the surface of our planet. These rays are germicidal. They are emitted by ultraviolet lamps commonly used to sterilize surgical instruments. Strong ultraviolet rays from the Sun are an extremely serious hazard to most forms of life on Earth. But this is because most forms of life on Earth evolved in the absence of a high ultraviolet flux.

It is easy to imagine adaptations to protect organisms against ultraviolet light. In fact, sunburn and high melanin pigmentation in the skin are adaptations in this direction. They have not been carried very far in most terrestrial organisms because the present ultraviolet flux is not very high. In a place like Mars, where there is little ozone, the ultraviolet light at the surface is extremely intense. But the Martian surface material is a strong absorber of ultraviolet light—as most soil and rocks are—and we can easily imagine organisms walking around with small ultraviolet-opaque shields on their backs: Martian turtles. Or perhaps Martian organisms carry about ultraviolet parasols. Many organic molecules also could be used in the exterior layers of extraterrestrial organisms to protect them against ultraviolet light.

There are temperature chauvinists. It is said that the freezing temperatures on planets like Jupiter or Saturn, in the outer Solar System, make all life there impossible. But these low temperatures do not apply to all portions of the planet. They refer only to the outermost cloud layers—the layers that are accessible to infrared telescopes that can measure temperatures. Indeed, if we had such a telescope in the vicinity of Jupiter and pointed it at Earth, we would deduce very low temperatures on Earth: We would be measuring the temperatures in the upper clouds and not on the much warmer surface of Earth.

It is now quite firmly established, both from theory and from radio observations of these planets, that as we penetrate below the visible clouds, the temperatures increase. There is always a region in the atmospheres of Jupiter, Saturn, Uranus, and Neptune that is at quite comfortable temperatures by terrestrial standards.

But why is it necessary to have temperatures like those on Earth in order for life to proliferate? A human being is seriously inconvenienced if his body temperature is raised or lowered by a mere 20 degrees. Is this because we happen to live by accident on the one planet in the Solar System that has a surface at the right temperature for biology? Or is it that our chemistry is delicately attuned to the temperature of the planet on which we have evolved? The latter is almost surely the case. Other temperatures, other biochemistries.

Our biological molecules are put together in complex three-dimensional arrangements. The functioning of these molecules, particularly the enzymes, are turned on and off by altering these three-dimensional arrangements. The chemical bonds that do these rearrangements must be weak enough to be broken conveniently at terrestrial temperatures, and at the same time strong enough not to fall to pieces if left alone for short periods of time. A chemical bond known as the hydrogen bond has an energy appropriately intermediate between these unreactive and unstable alternatives. The hydrogen bond is intimately connected with the three-dimensional biochemistry of terrestrial organisms.

On a much hotter planet like Venus, our biological molecules would fall to pieces. On a much colder planet in the outer Solar System, our biological molecules would be rigid, and our chemical reactions would not proceed at any useful rate. However, it is conceivable that much stronger bonds on Venus and much weaker bonds in the outer Solar System play the same role that hydrogen bonds play on Earth. We may have been much too quick to reject life at temperatures very different from those on our planet. There are not many chemical reactions known that can proceed at useful rates at some very low temperature such as might exist on Pluto, 30 or 40 degrees above absolute zero. But there are also very few chemical laboratories on Earth where experiments are performed at 30 or 40 degrees above absolute zero. With a few exceptions, such experiments have not been performed at all.

We are thus at the mercy of observational selection. We examine only a small fraction of the possible range of cases because of some unconscious bias, or the fact that scientists wish to work in their shirtsleeves. We then conclude that all conceivable cases must conform to what our preconceptions have forced upon us.

Another common chauvinism—one which, try as I might, I find I share—is carbon chauvinism. A carbon chauvinist holds that biological systems elsewhere in the universe will be constructed out of carbon compounds, as is life on this planet. There are conceivable alternatives: Atoms like silicon or germanium can enter into some of the same kinds of chemical reactions as carbon does. It is also true that much more attention has been paid to carbon organic chemistry than to silicon or germanium organic chemistry, largely because most biochemists we know are of the carbon, rather than the silicon or germanium, variety. Nevertheless, from what we know of the alternative chemistries, it appears clear that—except in very low-temperature environments—there is a much wider variety of complex compounds that can be built from carbon than from the alternatives.

In addition, the cosmic abundance of carbon exceeds that of silicon, germanium, or other alternatives. Everywhere in the universe, and particularly in primitive planetary environments in which the origin of life occurs, there is simply more carbon than alternative atoms available to make complex molecules. We see from laboratory experiments simulating the primitive atmosphere of the Earth or the present environment of Jupiter, as well as in radioastronomical studies of the interstellar medium, a profusion of simple and complex organic molecules readily produced by a wide variety of energy sources. For example, in one of our experiments, the passage of a single high-pressure shock wave through a mixture of methane (CH_4), ethane (C_2H_6), ammonia (NH_3), and water (H_2O) converted 38 percent of the ammonia into amino acids, the building blocks of proteins. There were not enormous quantities of other sorts of organic molecules.

Thus, both the atoms and the simple molecules of which we are made are probably common to organisms elsewhere in the universe. But the specific way in which these molecules are put together and the specific forms and physiologies of the extraterrestrial organisms may be, because of their different evolutionary histories, extremely different from what is common on our planet.

In considering which stars to examine for possible radio signals directed at us, much attention is usually given to stars like our Sun. It has been reasonably argued that searches should begin with the one type of star we know has life on at least one of its planets, namely stars like our own Sun. In Project Ozma, the first attempt to search for such radio signals, the two stars examined, Tau Ceti and Epsilon Eridani, were both stars with mass, radius, age, and composition very similar to our Sun, which astronomers call a G-0 dwarf. They were, in fact, the nearest two Sun-like stars.

But should we restrict our attention to stars like the Sun? I think not. Stars of slightly smaller mass and of slightly lower luminosity than our Sun are longer lived. These stars, called K and M dwarfs, can be many billions of years older than the Sun. If we imagine that the longer the lifetime of a planet, the more likely it is that intelligent organisms have evolved on it, we should then bias our searches toward K and M stars and avoid G-star chauvinism. It may be objected that planets of K and M stars are much colder than Earth, and that life on them may be less likely. The premise of this objection does not appear to be true; such planets seem to be closer to their stars than the corresponding planets in our Solar System; and we have already discussed the fallacies of temperature chauvinism. Too, there are many more K and M stars than G stars.

Is there planetary chauvinism? Must life arise and reside on planets, or might there be organisms that inhabit the depths of interstellar space, the surfaces or interiors of stars, or other even more exotic cosmic objects?

In our present state of ignorance, these are very difficult questions to answer. The density of matter in interstellar space is so low that an organism there simply cannot acquire enough material to make a copy of itself in any reasonable period of time. This is not true in dense interstellar clouds, but such clouds live for very short periods of time, condensing to form stars and planets. In the process, they become so hot that any organic compounds contained within them are probably destroyed.

We might imagine organisms evolving on planets with atmospheres slowly leaking away to space, permitting the organisms gradually to adapt to the increasingly severe conditions, and finally acclimatizing to what in effect is an interstellar environment. Organisms leaving such planets—perhaps by electromagnetic radiation pressure, or by solar wind from the local sun—might populate interstellar space, but they would still be faced with insurmountable problems of malnutrition.

A quite different sort of interstellar organism may be much more likely: Intelligent beings who arise on planets as we have, but who have moved their arena of activities to the much vaster volume of interstellar space. Beings in our far technological future should have capabilities at which we cannot today even dimly guess. It is not out of the question that such societies could tap the matter and energy of stars and galaxies for their own uses. Just as we are organisms completely at home only on the land, although we evolved from the sea, the universe may be populated with societies that arose on planets but that are comfortable only in the depths of interstellar space.

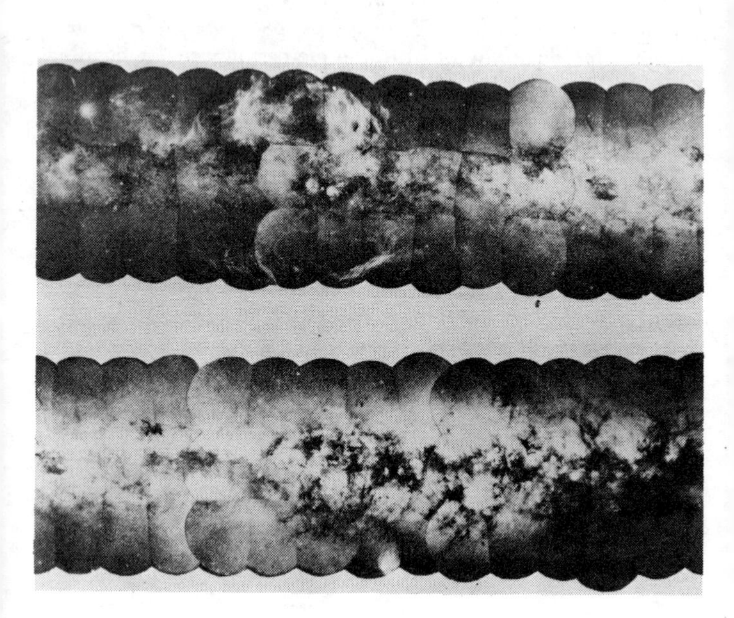

Composite photograph by all-sky cameras of our Milky Way Galaxy. Courtesy, Dr. Bart Bok and Lund Observatory.

7.

Space Exploration as a Human Enterprise

I. THE SCIENTIFIC INTEREST

There is a place with four suns in the sky—red, white, blue, and yellow; two of them are so close together that they touch, and star-stuff flows between them.

I know of a world with a million moons.

I know of a sun the size of the Earth—and made of diamond.

There are atomic nuclei a mile across that rotate thirty times a second.

There are tiny grains between the stars, with the size and atomic composition of bacteria.

There are stars leaving the Milky Way. There are immense gas clouds falling into the Milky Way.

There are turbulent plasmas writhing with X- and gamma-rays and mighty stellar explosions.

There are, perhaps, places outside our universe.

The universe is vast and awesome, and for the first time we are becoming a part of it.

The planets are no longer wandering lights in the evening sky. For centuries, Man lived in a universe that seemed safe and cozy—even tidy. Earth was the cynosure of creation and Man the pinnacle of mortal life. But these quaint and comforting notions have not stood the test of time. We now know that we live on a tiny clod of rock and metal, a planet smaller than some relatively minor features in the clouds of Jupiter and inconsiderable when compared with a modest sunspot.

Our star, the Sun, is small and cool and unprepossessing, one of some two hundred billion suns that make up the Milky Way Galaxy. We are located so far from the center of the

Milky Way that it takes light, traveling at 186,000 miles a second, some 30,000 years to reach us from there. We are in the galactic boondocks, where the action isn't. The Milky Way Galaxy is entirely unremarkable, one of billions of other galaxies strewn through the vastness of space.

No longer does "the world" mean "the universe." We live on one world among an immensity of others.

Charles Darwin's insights into natural selection have shown that there are no evolutionary pathways leading unerringly from simple forms to Man; rather, evolution proceeds by fits and starts, and most life forms lead to evolutionary dead-ends. We are the products of a long series of biological accidents. In the cosmic perspective there is no reason to think that we are the first or the last or the best.

These realizations of the Copernican and Darwinian revolutions are profound—and, to some, disturbing. But they bring with them compensatory insights. We realize our deep connectedness with other life forms, both simple and complex. We know that the atoms that make us up were synthesized in the interiors of previous generations of dying stars. We are aware of our deep connection, both in form and in matter, with the rest of the universe. The cosmos revealed to us by the new advances in astronomy and biology is far grander and more awesome than the tidy world of our ancestors. And we are becoming a part of it, the cosmos as it is, not the cosmos of our desires.

Mankind now stands at several historical branching points. We are on the threshold of a preliminary reconnaissance of the cosmos. For the first time in his history, Man is capable of sending his instruments and himself from his home planet to explore the universe around him.

But the exploration of space has been defended largely in terms of narrow considerations of national prestige, both in the United States and in the Soviet Union; in terms of the development of technological capabilities, in an age when many people are finding the development of technology for its own sake to have disastrous consequences; in terms of technological "spinoff" when the space program costs very much more than the cost of direct development of the spin-off; and in terms of a quite tenuous argument for military advantage, in a time when people the world over long for a demilitarization of society.

Under these circumstances, it is not surprising that hard questions are being asked about expenditures in space, when there are visible and urgent needs for funds to correct injustices and improve society and the quality of life on Earth. These questions are entirely appropriate. If scientists cannot give to the man on the street a satisfactory explanation of expenditures in the exploration of space, it is not obvious that public funds should be allocated for such ventures.

The interest of an individual scientist in space exploration is likely to be very personal—something puzzles him, intrigues him, has implications that excite him. But we cannot ask the public to spend large sums just to satisfy the scientist's curiosity. When we probe more deeply into the professional interests of individual scientists, however, we often find a focus of concern that largely overlaps the public interest.

A fundamental area of common interest is the problem of perspective. The exploration of space permits us to see our planet and ourselves in a new light. We are like linguists on an isolated island where only one language is spoken. We can construct general theories of language, but we have only one example to examine. It is unlikely that our understanding of language will have the generality that a mature science of human linguistics requires.

There are many branches of science where our knowledge is similarly provincial and parochial, restricted to a single example among a vast multitude of possible cases. Only by examining the range of cases available elsewhere can a broad and general science be devised.

The science that has by far the most to gain from planetary exploration is biology. In a very fundamental sense, biologists have been studying only one form of life on Earth. Despite the apparent diversity of terrestrial life forms, they are identical in the deepest sense. Beagles and begonias, bacteria and baleen whales all use nucleic acids for storage and transmission of hereditary information. They all use proteins for catalysis and control. All organisms on Earth, so far as we know, use the same genetic code. The cross-sectional structures of human sperm cells are almost identical with those of the cilia of paramecia. Chlorophyll and hemoglobin and the substances responsible for the coloring of many animals are all essentially the same molecule.

It is difficult to escape the conclusion—which, in a sense, is implicit in Darwinian natural selection—that all life on Earth has evolved from a single instance of the origin of life. If this is true, there is an important sense in which the biologist cannot distinguish the necessary from the contingent, that is, distinguish those aspects of life that any organism anywhere in the universe must have simply in order to be alive, from those aspects of life that are the results of the tortuous evolution by small opportunistic adaptations.

The production of simple organic (carbon-based) molecules under simulated primitive planetary conditions is now subject to active laboratory investigation. As we saw earlier, the molecules of which we are made can be produced rather easily, in the absence of life, under quite general primitive planetary conditions. But it is not practicable to perform laboratory experiments on even the early stages of biological evolution: The time scales are too long. It is only by examining living systems elsewhere that biologists can determine what other possibilities there are.

It is for this reason that the discovery of even an extremely simple organism on Mars would have profound biological significance. On the other hand, if Mars proves to be lifeless, a natural experiment has been performed for us: Two planets, in many respects similar; but on one life has evolved, on the other it has not. By comparing the experimental with the control planet, much may be discovered about the origin of life. Similarly, the search for prebiological organic chemicals on the Moon, on Mars, or on Jupiter is of great importance in understanding the steps leading to the origin of life.

As another example of the perspective provided by planetary studies, consider meteorology. The problems of turbulent flow and fluid dynamics are among the most difficult in all of physics. Some insights into the Earth's weather have been obtained by standing back and examining, by photography from meteorological satellites, the circulation of the Earth's atmosphere. Still, meteorological theory for the Earth is today capable of long-range weather predictions, but only over very large geographical areas, for a range of simplifying assumptions, and for only a little time into the future. Laboratory studies of atmospheric circulation have a limited scope; classically, they are performed in modified dishpans.

It would be nice to do a "Joshua" experiment, stopping the Earth from turning for a while. The change in circulation would provide insights into the role of the Earth's rotation (particularly through Coriolis forces) in determining the circulation. But such an experiment is technologically very difficult. It also has undesirable side-effects. On the other hand, the planet Venus, with approximately the same mass and radius as Earth, has a rotation rate 240 times slower—so slow that Coriolis forces will be minor. The atmosphere is much thicker on Venus than on Earth. Nature has arranged a natural experiment for the meteorologists.

Jupiter rotates about once every ten hours; here is an enormous planet that turns faster than Earth does. The effects of rotation should be much more important than on Earth, and, indeed, Jupiter gives the impression of having a seething, roiling, turbulent atmosphere; its prominent atmospheric bands and belts are almost certainly related to the rapid rotation. Nature has arranged two comparison experiments—two planets with massive atmospheres, one rotating slowly, the other rapidly. An understanding of the circulation of the massive atmospheres of Venus and Jupiter will improve our understanding of oceanic, as well as atmospheric, circulation on Earth.

Or consider the planet Mars. Here is a planet with—quite remarkably—the same period of rotation and the same inclination of its axis of rotation to its orbital plane as Earth. But its atmosphere is only 1 percent of ours, and it has no oceans and no liquid water. Mars is a control experiment on the influence of oceans and liquid water on atmospheric circulation.

Until recently, the geologist has been restricted to one object of study, the Earth. He was unable to decide which properties of the Earth are fundamental to all planetary surfaces and which are peculiar to the unique circumstances of Earth. For example, seismographic observations of earthquakes have revealed the interior structure of Earth and its division into crust, mantle, liquid metal core, and solid inner core. But the reason Earth is so divided remains largely obscure. Was Earth's crust exuded from the mantle through geological time? Did it fall from the skies in an early catastrophic event? Has Earth's core formed gradually through geological time by the sinking of iron through the mantle? Or did it form discontinuously, perhaps in a molten Earth at the time of the origin of our planet? Such questions can be examined by performing seismometric observations on the surfaces of other planets; they could be relatively inexpensive experiments performed automatically by existing instrumentation.

There is now reasonably convincing evidence of continental drift. The motion of Africa and South America away from each other is the best-known example. In some theories, the driving force of continental drift and of the evolution of the interior of our planet are connected—for example, through convection currents circulating slowly between core and crust in the mantle. Such connections between surface geology and planetary interiors are just beginning to emerge in the study of other planets. We test our understanding of such connections by testing whether they apply elsewhere.

The perspectives gained in studies like these have a range of practical consequences. A generalization of the science of meteorology may lead to great improvements in weather forecasting. It may even lead to weather modification. The study of the atmosphere of Venus has already led to the theory that a runaway "greenhouse" effect has occurred there—an unstable equilibrium in which an increase in temperature leads to an increase in the atmospheric water vapor content, leading through infrared absorption of thermal radiation from the planet to a yet further increase in surface temperature, and so on. Had Earth started out only slightly closer to the Sun than it did, preliminary theoretical estimates indicate that we might have ended up as a searing hot Venus. But we live in a time when the atmosphere of Earth is being strongly modified by the activities of Man. It is of the first importance to understand precisely what happened on Venus so that an accidental recapitulation on Earth of the runaway Venus greenhouse can be avoided.

The studies of the surfaces and interiors of the planets may be of great practical benefit in earthquake prediction and in remote geological prospecting for minerals of value on Earth.

The revolution in biology that the discovery of indigenous life elsewhere would surely bring may have a range of unsuspected practical benefits, particularly to the extent that research in cancer and aging is now limited by ideas rather than money.

The study of the highly condensed matter in neutron stars and the enormous energy productions in the centers of galaxies and in quasars has already led to suggestions about possible modifications of the laws of physics, laws that have been deduced on Earth to explain phenomena observed on Earth.

The exploration of space will inevitably provide a wealth of practical benefits. But the history of science suggests that the most important of these will be unexpected—benefits we are today not wise enough to anticipate.

Human footprints on the Moon. The party is over and the guests have left. The footprints were made by the *Apollo 15* crew and will last for a million years. Mount Hadley is in the distance.

8.

Space Exploration as a Human Enterprise

II. THE PUBLIC INTEREST

Direct scientific interest in space exploration and the practical consequences that can be imagined flowing from them are not the principal or even the most general interests that space exploration holds for the layman. There is today—in a time when old beliefs are withering—a kind of philosophical hunger, a need to know who we are and how we got here. There is an ongoing search, often unconscious, for a cosmic perspective for humanity. This can be seen in innumerable ways, but most clearly on the college campus. There, an enormous interest is apparent in a range of pseudoscientific or borderline-scientific topics—astrology, scientology, the study of unidentified flying objects, investigation of the works of Immanuel Velikovsky, and even science-fiction superheroes—all of which represent an attempt, overwhelmingly unsuccessful in my view, to provide a cosmic perspective for mankind. Professor George Wald, of Harvard, is thinking of this longing for a cosmic perspective when he writes: "We have desperately to find our way back to human values. I would even say to religion. There is nothing supernatural in my mind. Nature is my religion, and it's enough for me. . . . What I mean is: We need some widely shared view of the place of Man in the Universe."

The most widely sold book in college communities from Cambridge, Massachusetts, to Berkeley, California, in recent years was called *The Whole Earth Catalog*, which viewed itself as providing access to tools for the creation of cultural alternatives. What was striking was the number of works displayed in the *Catalog* that related to a scientific cosmic perspective. They ranged from the Hubble *Atlas of Galaxies* to flags and posters of Earth photography near full phase. The title of *The Whole Earth Catalog* derives from its founder's urge to see a photograph of our planet as a whole. The Fall 1970 issue expanded this perspective, showing a photograph of the whole Milky Way Galaxy.

There is a similar trend apparent in some modern art and in rock 'n' roll music: "Cosmo's Factory" by Creedence Clearwater Revival, "Starship" by the Jefferson Airplane, "Mr. Spaceman" and "CTA 102" by the Byrds, "Mr. Rocket Man" by Elton John, and many others.

Such interest is not restricted to the young. There is a tradition in the United States of public-subscription support of astronomy. Construction of entire observatories and salaries for staff have been paid for voluntarily by the local communities. Several million people visit planetariums in North America and Britain each year.

The current resurgence of interest in the ecology of the planet Earth is also connected with this longing for a cosmic perspective. Many of the leaders of the ecological movement in the United States were originally stimulated to action by photographs of Earth taken from space, pictures revealing a tiny, delicate, and fragile world, exquisitely sensitive to the depredations of man—a meadow in the middle of the sky.

As the results of space exploration and their accompanying new perspectives on Earth and its inhabitants permeate our society, they must, I believe, have consequences in literature and poetry, in the visual arts and music. The distinguished American physicist Richard Feynman writes: "It does no harm to the mystery to know a little about it. For far more marvelous is the truth than any artists of the past imagined! Why do the poets of the present not speak of it? What men are poets who can speak of Jupiter if he were like a man, but if he is an immense spinning sphere of methane and ammonia must be silent?"[1]

But mere general exploration does not yet motivate pervasive public interest in space. For many, the rocks returned

[1] Richard Feynman, *Introduction to Physics,* Vol. I, Addison Wesley, pp. 3–6.

from the Moon were a great disappointment. They were seen as merely rocks. What role they might play in chronicling the days of creation of the Earth-Moon system has not yet been explained adequately to the public.

Where public interest in space is most apparent is in cosmology and in the search for extraterrestrial life, topics that strike resonant chords in a significant fraction of mankind. The fact that much more newspaper space is given to the most casual hypothesis on exobiology than to many of the most careful and important results in other areas is an accurate reflection of where public interest lies. The discovery of interstellar microwave lines of formaldehyde and hydrogen cyanide has been widely described in the public press as connected, through a long set of linkages, to questions in biology.

While it is true that the average person thinks in terms of mild variants of human beings when he is asked to imagine extraterrestrial life, it is also true that interest even in Martian microbes is much larger than in many other areas of space exploration. The search for extraterrestrial life could be a keystone of public support for space experiments—experiments oriented both within and beyond the Solar System.

There are many possible viewpoints on the present and near-future costs of space science and astronomy. Because the annual costs of ground-based astronomy are only a few percent of the costs of the scientific space program, I will concentrate on the price of the latter. It is customary to compare expenditures on space to annual expenditures in the United States for ethyl alcohol or bubble gum or cosmetics. I personally find it more useful to compare the costs with those of the U. S. Department of Defense. Using a report of the government's General Accounting Office (the New York *Times*, July 19, 1970), we learn that the total anticipated cost of the Viking mission to land on Mars in 1976 is about half that of the cost *overruns* in the so-called Safeguard antiballistic missile system for fiscal year 1970. The cost of a Grand Tour exploration of all the planets in the outer Solar System (canceled for lack of funds) is comparable to the 1970 cost *overruns* on the Minuteman III system; the cost of a very large optical telescope in space, capable of definitive studies of the origins of the universe, is comparable to the 1970 cost *overruns* on the Minuteman II missile; and a major program of Earth resources satellites, involving several years of close inspection of the surface and weather of our planet, would cost approximately the fiscal year 1970 cost *overruns* on the P-3C aircraft. A decade-long program of systematic investigation of the entire Solar System would cost as much as the accounting mistakes on a single "defense" weapons system in a single year. The scientific space program is small change compared to the *errors* in the Department of Defense budget.

Another viewpoint worth considering is space exploration as entertainment. A Viking Mars-lander could be completely funded through the sale, to every American, of a single issue of a magazine, containing pictures taken on the surface of Mars by Viking. Photographs of the Earth, the Moon, the planets, and spiral and irregular galaxies are an appropriate and even characteristic art form of our age. Such novel and oddly moving photographs as the Lunar Orbiter image of the interior of the crater Copernicus and the *Mariner 9* photography of the Martian volcanoes, windstreaks, moons, and polar icecaps speak both to a sense of wonder and to a sense of art. An unmanned roving vehicle on Mars could probably be supported by subscription television. A phonograph record of the output of a microphone on Mars, where there seems to be a great deal of acoustic energy, might have wide sales.

I do not wish here to broach the debate on manned vs. unmanned planetary exploration, except to stress again that there may be very good nonscientific reasons for sending men into space. There exist intermediate cases between manned and unmanned exploration, which we may very well see in the forthcoming decades. For example, there may be telepuppets, devices landed on another planet but fully controlled by an individual human being in orbit, all of whose senses are in a feedback loop with the device. It is also possible that planets with very hostile environments by terrestrial standards, or planets where there is a great danger of contamination by terrestrial micro-organisms, will be explored by men inside machines like enormous prosthetic devices, amplifying the sense perceptions and muscular abilities of the human operator.

Even apart from these hypothetical developments, it is already quite clear that the development of sophisticated devices for unmanned planetary exploration is organizing the same technology required for the production of useful robots on the Earth. An unmanned vehicle that lands on Mars by the early 1980s will very likely have the ability to sense its environment much more thoroughly than humans are able to, to rove over the landscape, to make both preprogrammed decisions and decisions based on information newly acquired. First cousins to such a robot, some mass-produced, would be extremely useful devices here on Earth. I am thinking in part of operations in inaccessible environments such as the abyssal floor of the ocean basins, but I am also thinking of industrial robots to free workers from repetitive and uninteresting tasks, and domestic robots to liberate the housekeeper from a life of drudgery.

The experience of space exploration gives no unique philosophy; to some extent, each group tends to see its own philosophical view reflected, and not always by the soundest logic: Nikita Khrushchev stressed that in the space flight of Yuri Gagarin no angels or other supernatural beings were detected; and, in almost perfect counterpoint, the *Apollo 8* astronauts read from lunar orbit the Babylonian cosmogony enshrined in Genesis, Chapter 1, as if to reassure their American audience that the exploration of the Moon was not really in contradiction to anyone's religious beliefs. But it is striking how space exploration leads directly to religious and philosophical questions.

I believe that military control of manned space flight—in practice in the Soviet Union and a subject of current debate in the United States—is a step that supporters of peace should back. The military establishments of the United States and the Soviet Union are, I am afraid, establishments with vested interests in war. They are meticulously trained for war; in time of war, there are rapid promotions, increases in pay, and opportunities for valor that are absent in peacetime. Where eager readiness for warfare exists, the likelihood of intentional or accidental warfare becomes much greater. By virtue of their training and temperament, military men are often not interested in other sorts of gainful employment. There are few other ways of life with the perquisites of power of the military officer. If peace broke out, the officer corps, their services no longer as necessary, would be profoundly discomfited. Premier Khrushchev once attempted to cashier a large number of senior officers in the Red Army, putting them in charge of hydroelectric power stations and the like. This was not to their liking, and in something like a year most of them were back in their old jobs. In fact, the military establishments in the United States and the Soviet Union owe their jobs to each other, and there is a very real sense in which they form a natural alliance against the rest of us.

At the same time, there are enormous labor forces and huge electronics, missile, and chemical industries that have an equally strong vested interest in and maintain equally strong lobbies for the maintenance of the warfare state. Barring some awesomely atypical epidemic of reason, is there not some way that this powerful collection of vested interests could be moved toward more peaceful activities? Space exploration requires exactly this combination of talents and capabilities. It requires a large technical base in such areas as electronics, computer technology, precision machinery, and aerospace frames. It requires something very close to military organization to keep a large number of geographically dispersed enterprises moving in phase toward a common goal.

The history of the exploration of the Earth's surface has largely been a military history, in part because it is an appropriate application of military traditions of organization and personal valor. It is the other military traditions that pose a danger to us today. Perhaps the exploration of the Solar System is an alternative and honorable employment for the military and industrial vested interests. I can imagine a transition to an arrangement where a significant fraction of the career officer corps of both the United States and the U.S.S.R. is transferred to space exploration. At least in part because of their considerable abilities, a fair number of military officers are employed by the National Aeronautics and Space Administration in activities with little or no military significance. And of course the vast majority of astronauts and cosmonauts have been military officers. This is surely all to the good: The more of them engaged up there, the less of them engaged down here.

Despite many NASA press releases, much of the space program and almost all of its space science and applications program are in the long run not gimmicky or parochial in perspective. Indeed, they share a community of philosophical, exploratory, and human interest with many segments of American society—even segments that are in strong mutual disagreement on many issues. The cost of space exploration seems very modest compared with its potential returns.

9.

Space Exploration as a Human Enterprise

III. THE HISTORICAL INTEREST

In the long view, the greatest significance of space exploration is that it will irreversibly alter history. As we mentioned in Chapter 1, the group with which Man identifies has gradually broadened during this history of mankind. Today the bulk of the world's population has at least a major personal identification with national superstates. While progress has not been smooth, and there are occasional reversals, the trend is clearly toward a group identification with mankind as a whole. Space exploration can hasten this identification. Astronauts and cosmonauts have remarked with great feeling about the beauty and serenity of the Earth viewed from space. For many of them, a flight into space has been a religious experience, transfiguring their lives. National boundaries do not appear in photographs of Earth from space. As Arthur C. Clarke somewhere remarked, it is difficult to imagine even the most fervent of nationalists not reconsidering his views as he sees the Earth fade from a faint crescent to a tiny point of light, lost among millions of stars.

Space exploration provides a calibration of the significance of our tiny planet, lost in a vast and unknown universe. The search for life elsewhere will almost surely drive home the uniqueness of Man: The winding, unsure, improbable, evolutionary pathway that has brought us to where we are; and the improbability of finding—even in a universe populated with other intelligences—one with a form very much like our own. In this perspective, the similarities among men will stand out overwhelmingly against our differences.

There is a practical geocentrism to our everyday life. We still talk about the Sun rising and setting rather than the

Earth turning. We still think of the universe organized for our benefit and populated only by us. Space exploration will bring also a little humility.

Harold Urey has perceptively referred to the space program as a kind of contemporary pyramid-building. Seen in the context of Pharaonic Egypt, the analogy seems particularly apt, for the pyramids were an attempt to deal with problems of cosmology and immortality. In the long historical perspective, this is precisely what the space program is about. The footprints left by astronauts on the Moon will survive a million years, and the miscellaneous instruments and packing cases left there may last as long as the Sun.

On the other hand, the pyramids are monumental and, we today believe, futile efforts to insure the survival after death of one man, the Pharaoh. Perhaps a better analogy is with the ziggurats, the terraced towers of the Sumerians and Babylonians—the places where the gods came down to Earth and the population as a whole transcended everyday life. There is no doubt a little of the pyramid in the great rocket boosters; but I think their ultimate significance is more likely to be as contemporary ziggurats.

A society engaged in a relatively modest, peaceful, and intellectually significant exploration of its surroundings garners thereby the possibility of achieving greatness. It is difficult to prove such causal chains, and, historically, there are no one-to-one correlations. But it is remarkable that the nations and epochs marked by the greatest flowering of exploration are also marked by the greatest cultural exuberance. In part, this must be because of the contact with new things, new ways of life, and new modes of thought unknown to a closed culture, with its vast energies turned inward.

There are examples from the Biblical Near East, from Periclean Athens, and from other times, but I am most taken by the example of the age of European exploration and discovery. The vernacular languages of France, England, and Iberia found a definitive literary expression at the same time that the earliest transatlantic voyages of discovery were occurring. Rabelais and Montaigne in France; Shakespeare,

Milton, and the translators of the King James Bible in England; Cervantes and Lope de Vega in Spain; Camoens in Portugal, all date from this period. From the writings of Francis Bacon it is clear that exposure to new parts of the world had a profound influence on the thinking of the times. This period saw the invention of such fundamental instruments as the telescope, the microscope, the thermometer, the barometer, and the pendulum clock.

It was also the epoch of Galileo (1564–1642), who, while not resident in one of the new exploratory nations, was closely tied to one of them—Holland, where the telescope that he improved upon was first invented. Many of the works of graphic art during this period—for example, those of Hieronymus Bosch and El Greco—reflect the spirit of change that permeated the times. It was the era of the establishment of modern physics by Isaac Newton. Descartes, Hobbes, and Spinoza—pivotal individuals in the history of philosophy—flourished. In the activities and writings of da Vinci, Gilbert, Galileo, and Bacon, the period also corresponds to the origin of the experimental method in science.

An interesting case history is provided by Holland, a country that has provided more than its fair share of men of learning and culture. If there was one moment of cultural efflorescence in Holland, it was the period centered around the last half of the seventeenth century. Iberian ports were inaccessible to the Dutch Republic because of the war between France and Spain. Forced to find its own sources of trade, Holland founded the Dutch East and West India Companies. A significant fraction of the national resources was put into seafaring; one consequence was that Holland became—for the only time in its history—a world power. Because of these ventures, Dutch is spoken in Indonesia today, and several individuals of Dutch ancestry rose to the Presidency of the United States. Far more important is the fact that, during the same period, Vermeer and Rembrandt, Spinoza and van Leeuwenhoek flourished in Holland. It was a tightly knit society: Van Leeuwenhoek, was, in fact, the executor of Vermeer's estate. Holland was the most liberal and least authoritarian nation in Europe during this time.

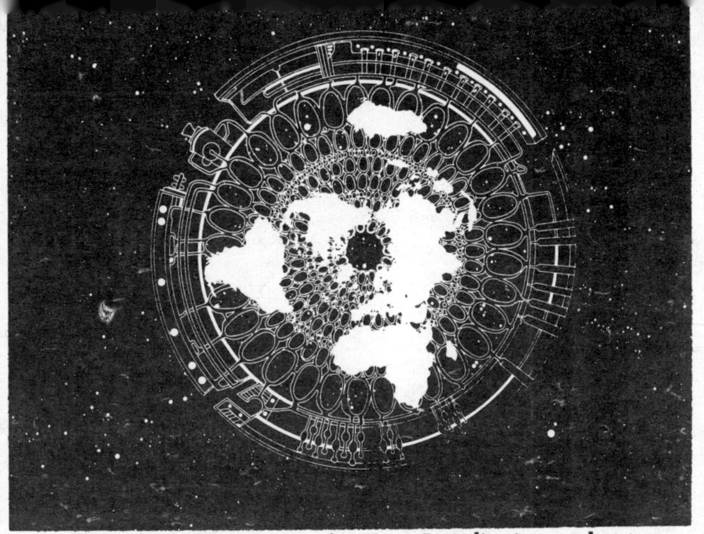

Schematic representation of a Type I civilization, perhaps a few centuries more advanced than our civilization. By Jon Lomberg.

In all the history of mankind, there will be only one generation that will be first to explore the Solar System, one generation for which, in childhood, the planets are distant and indistinct discs moving through the night sky, and for which, in old age, the planets are places, diverse new worlds in the course of exploration.

There will be a time in our future history when the Solar System will be explored and inhabited. To them, and to all who come after us, the present moment will be a pivotal instant in the history of mankind. There are not many generations given an opportunity as historically significant as this one. The opportunity is ours, if we but grasp it. To paraphrase K. E. Tsiolkovsky, the founder of astronautics: The Earth is the cradle of mankind, but one cannot live in the cradle forever.

A human infant begins to achieve maturity by the experimental discovery that he is not the whole of the universe. The same is true of societies engaged in the exploration of their surroundings. The perspective carried by space exploration may hasten the maturation of mankind—a maturation that cannot come too soon.

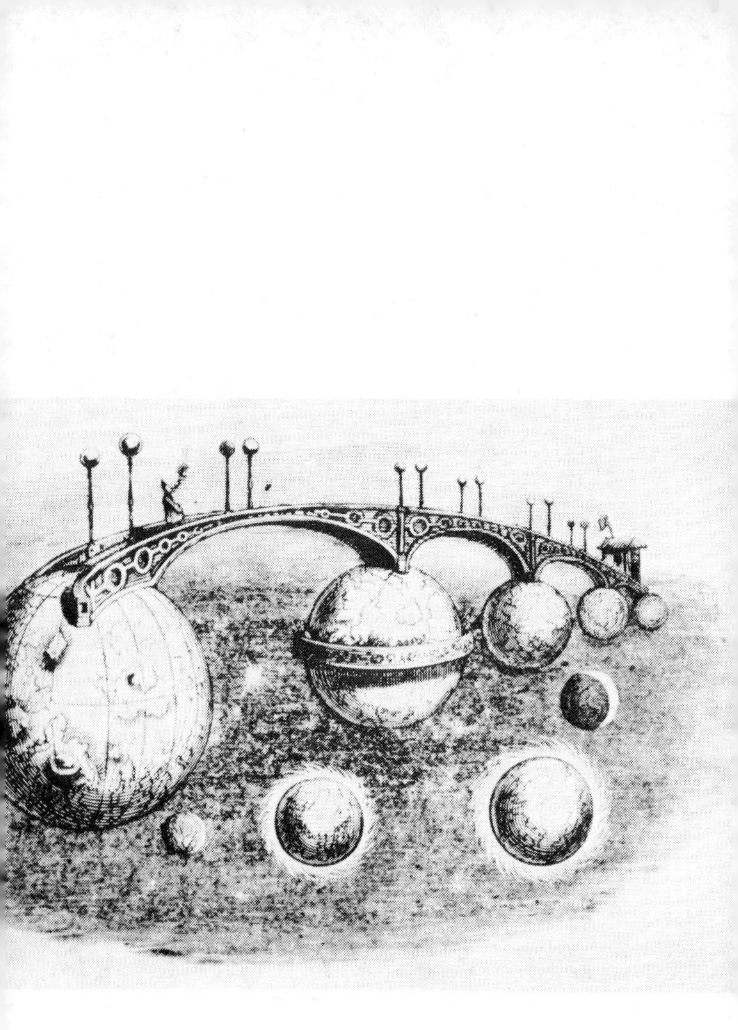

"Les Mystères des Infinis" by Grandville, 1844.

Part Two

THE SOLAR SYSTEM

There was a time—and very recently—when the idea of the possibility of learning the composition of the celestial bodies was considered senseless even by prominent scientists and thinkers. That time has now passed. The idea of the possibility of a closer, direct, study of the universe will today, I believe, appear still wilder. To step out onto the soil of asteroids, to lift with your hand a stone on the moon, to set up moving stations in ethereal space, and establish living rings around the earth, the moon, the sun, to observe Mars from a distance of several tens of versts, to land on its satellites and even on the surface of Mars—what could be more extravagant! However, it is only with the advent of reactive vehicles that a new and great era in astronomy will begin, the epoch of a careful study of the sky. . . . The prime motive of my life is to do something useful for people. . . . That is why I have interested myself in things that did not give me bread or strength. But I hope that my studies will, perhaps soon but perhaps in the distant future, yield society mountains of grain and limitless power.

—K. E. Tsiolkovsky, 1912

Apollo photograph of the full Earth. Courtesy, NASA.

10.
On Teaching the First Grade

A friend in the first grade asked me to come to talk to his class, which, he assured me, knew nothing about astronomy but was eager to learn. With the approval of his teacher, I arrived at his school in Mill Valley, California, armed with twenty or thirty color slides of astronomical objects—the Earth from space, the Moon, the planets, exploding stars, gaseous nebulae, galaxies, and the like—which I thought would amaze and intrigue and, perhaps to a certain extent, even educate.

But before I began the slide show for these bright-eyed and cherubic little faces, I wanted to explain that there is a big difference between stating what science has discovered and describing how scientists found it all out. It is pretty easy to summarize the conclusions. It is hard to relate all the mistakes, false leads, ignored clues, dedication, hard work, and painful abandonment of earlier views that go into the initial discovery of something interesting.

I began by saying, "Now you have all *heard* that the Earth is round. Everybody *believes* that the Earth is round. But *why* do we believe the Earth is round? Can any of you think of any evidence that the Earth is round?"

For most of the history of mankind, it was reverently held that the Earth is flat—as is entirely obvious to anyone who has stood in a Nebraska cornfield around planting time. The concept of a flat Earth is still built into our language in such phrases as "the four corners of the Earth." I thought I would stump my little first-graders and then explain with what difficulty the sphericity of Earth had come into general human consciousness. But I had underestimated the first grade of Mill Valley.

"Well," asked a moppet in the sort of one-piece coverall worn by railroad engineers, "what about this business of a ship that's sailing away from you, and the last thing you see is the master, or whatever it's called, that holds up the sail? Doesn't that mean that the ocean has to be curved?"

"What about when there's an *ellipse* of the Moon? That's when the Sun is behind us and the shadow of the Earth is on the Moon, right? Well, I saw an *ellipse*. That shadow was round, it wasn't straight. So the Earth has to be round."

"There's better proof, much better proof," offered another. "What about that old guy who sailed around the world—Majello? You can't sail *around* the world if it isn't round, right? And people today sail around the world and fly around the world all the time. How can you fly around the world if it isn't round?"

"Hey, listen, you kids, don't you know there's *pictures* of the Earth?" added a fourth. "Astronauts have been in space, they took pictures of the Earth; you can look at the pictures, the pictures are all round. You don't have to use all these funny reasons. You can *see* that the Earth is round."

And then, as the *coup de grâce*, one pinafored little girl, recently taken on an outing to the San Francisco Museum of Science, casually inquired, "What about the Foucault pendulum experiment?"

It was a very sobered lecturer who went on to describe the findings of modern astronomy. These children were not the offspring of professional astronomers or college teachers or physicians or the like. They were apparently ordinary first-grade children. I very much hope—if they can survive twelve to twenty years of regimenting "education"—that they will hurry and grow up and start running things.

Astronomy is not taught in the public schools, at least in America. With a few notable exceptions, a student can pass from first to twelfth grade without ever encountering any of the findings or reasoning processes that tell us where we are in the universe, how we got here, and where we are likely to be going; without any confrontation with the cosmic perspective.

The ancient Greeks considered astronomy one of the half dozen or so subjects required for the education of free men. I find, in discussions with first-graders and hippie communards, congressmen and cab drivers, that there is an enormous untapped reservoir of interest and excitement in things astronomical. Most newspapers in America have a daily syndicated astrology column. How many have a daily syndicated astronomy column, or even a science column?

Astrology pretends to describe an influence that pervades people's lives. But it is a sham. Science really influences people's lives, and in only a slightly less direct sense. The enormous popularity of science fiction and of such movies as *2001: A Space Odyssey* is indicative of this unexploited scientific enthusiasm. To a very major extent, science and technology govern, mold, and control our lives—for good and for ill. We should make a better effort to learn something about them.

Britannia, the goddess on the obverse face of the old British penny.

11.
"The Ancient and Legendary Gods of Old"

The sorts of scientific problems that I am involved in—the environments of other planets, the origin of life, the possibility of life on other worlds—engage the popular interest. This is no accident. I think all human beings are excited about these fundamental problems, and I am lucky enough to be alive at a time when it is possible to perform scientific investigation of some of these problems.

One result of popular interest is that I receive a great deal of mail, all kinds of mail, some of it very pleasant, such as from the people who wrote poems and sonnets about the plaque on *Pioneer 10;* some of it from schoolchildren who wish me to write their weekly assignments for them; some from strangers who want to borrow money; some from individuals who wish me to check out their detailed plans for ray guns, time warps, spaceships, or perpetual motion machines; and some from advocates of various arcane disciplines such as astrology, ESP, UFO-contact stories, the speculative fiction of von Danniken, witchcraft, palmistry, phrenology, tea-leaf reading, Tarot cards, the I-Ching, transcendental meditation, and the psychedelic drug experience. Occasionally, also, there are sadder stories, such as from a woman who was talked to from her shower head by inhabitants of the planet Venus, or from a man who tried to file suit against the Atomic Energy Commission for tracking his every movement with "atomic rays." A number of people write that they can pick up extraterrestrial intelligent radio signals through the fillings in their teeth, or just by concentrating in the right way.

But over the years there is one letter that stands out in my mind as the most poignant and charming of its type. There came in the post an eighty-five-page handwritten letter, written in green ballpoint ink, from a gentleman in a mental hospital in Ottawa. He had read a report in a local newspaper that I had thought it possible that life exists on other planets; he wished to reassure me that I was entirely correct in this supposition, as he knew from his own personal knowledge.

To assist me in understanding the source of his knowledge, he thought I would like to learn a little of his personal history —which explains a good bit of the eighty-five pages. As a young man in Ottawa, near the outbreak of World War II, my correspondent chanced to come upon a recruiting poster for the American armed services, the one showing a goateed old codger pointing his index finger at your belly button and saying, "Uncle Sam Wants You." He was so struck by the kindly visage of gentle Uncle Sam that he determined to make his acquaintance immediately. My informant boarded a bus to California, apparently the most plausible habitation for Uncle Sam. Alighting at the depot, he inquired where Uncle Sam could be found. After some confusion about surnames, my informant was greeted by unpleasant stares. After several days of earnest inquiry, no one in California could explain to him the whereabouts of Uncle Sam.

He returned to Ottawa in a deep depression, having failed in his quest. But almost immediately, his life's work came to him in a flash. It was to find "the ancient and legendary gods of old," a phrase that reappears many times throughout the letter. He had the interesting and perceptive idea that gods survive only so long as they have worshipers. What happens then to the gods who are no longer believed in, the gods, for example, of ancient Greece and Rome? Well, he concluded, they are reduced to the status of ordinary human beings, no longer with the perquisites and powers of the godhead. They must now work for a living—like everyone else. He perceived that they might be somewhat secretive about their diminished circumstances, but would at times complain about having to do menial labor when once they supped at Olympus. Such retired deities, he reasoned, would be thrown into insane asylums. Therefore, the most reasonable method of locating these defrocked gods was to incarcerate himself in the local mental institution—which he promptly did.

While we may disagree with some of the steps in his reasoning, we probably all agree that the gentleman did the right thing.

My informant decided that to search for all the ancient and legendary gods of old would be too tiring a task. Instead, he set his sights on only a few: Jupiter, Mercury, and the goddess on the obverse face of the old British penny—not everyone's first choice of the most interesting gods, but surely a representative trio. To his (and my) astonishment, he found—incarcerated in the very asylum in which he had committed himself—Jupiter, Mercury, and the goddess on the obverse face of the old English penny. These gods readily admitted their identities and regaled him with stories of the days of yore when nectar and ambrosia flowed freely.

And then my correspondent succeeded beyond his hopes. One day, over a bowl of Bing cherries, he encountered "God Almighty," or at least a facsimile thereof. At least the Personage who offered him the Bing cherries modestly acknowledged being God Almighty. God Almighty luckily had a small spaceship on the grounds of the asylum and offered to take my informant on a short tool around the Solar System—which was no sooner said than done.

"And this, Dr. Sagan, is how I can assure you that the planets are inhabited."

The letter then concluded something as follows: "But all this business about life elsewhere is so much speculation and not worth the really serious interest of a scientist such as yourself. Why don't you address yourself to a really important problem, such as the construction of a trans-Canadian railroad at high northern latitudes?" There followed a detailed sketch of the proposed railway route and a standard expression of the sincerity of his good wishes.

Other than stating my serious intent to work on a trans-Canadian railroad at high northern latitudes, I have never been able to think of an appropriate response to this letter.

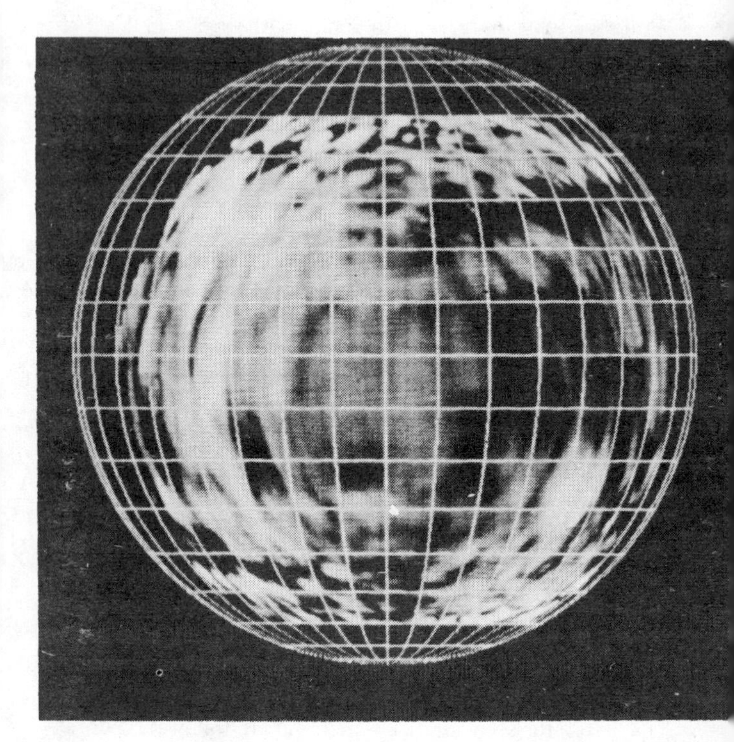

Radar map of the surface of Venus, unseen by the human eye because of the dense atmosphere and cloud cover of the planet. Courtesy, Arecibo Observatory, Cornell University.

12.
The Venus Detective Story

One of the reasons that planetary astronomy is such a delight these days is that it is possible to find out what's really right. In the old days, you could make any guess you liked, however improbable, about a planetary environment, and there was little chance that anyone could ever prove you wrong. Today, spacecraft hang like swords of Damocles over each hypothesis spun by planetary theoreticians, and the theoreticians can be observed in a curious amalgam of hope and fear as each new burst of spacecraft planetary information comes winging in.

Back when astronomers had telescopes, eyeballs, and very little else to assist their observations, Venus beckoned as a sister world. By the late nineteenth century, it was known that Venus had about the same mass and radius as Earth. Venus is the closest planet to Earth, and it was natural to assume that it was, in other respects, Earth-like.

Immanuel Kant imagined a race of amorous quasi-humans on Venus. Emanuel Swedenborg and Annie Besant, a founder of theosophy, found—by methods described as spirit travel and astral projection—creatures very like humans on Venus. In more recent years, some of the more spectacularly audacious flying-saucer accounts—for example, those of George Adamski—populated Venus with a race of benign and powerful beings, many of whom seem to have been garbed in long hair and long white robes—clear symbolism, in pre-1963 America, of deep spiritual intent. There is a long history of wishful thinking, bemused speculation, and conscious and unconscious fraud, which produced a popular expectation that our nearest planetary neighbor is habitable by humans, and is possibly even already inhabited by creatures rather like us.

It was, therefore, with a sense of considerable surprise, and even annoyance, that the results of the first radio observations of Venus were greeted. These measurements, performed in 1956 by C. H. Mayer and colleagues at the U. S. Naval Research Laboratory, found Venus to be a much more intense source of radio emission than had been expected. From Venus' distance to the Sun and the amount of sunlight it reflects back to space, the planet should be cool. Because Venus reflects so much sunlight back to space, its temperature ought to be even less than the Earth's, despite its closeness to the Sun. Mayer's group found that Venus, at a radio wavelength of 3 centimeters, was giving off as much radiation as it would if it were a hot body at a temperature of about 600 degrees Fahrenheit. Later observations with many different radio telescopes at many different radio frequencies confirmed the general conclusion that Venus had a "brightness temperature" of about 600 degrees to 800 degrees.

Nevertheless, there was great reluctance in the scientific community to believe that the radio emission came from the Venus surface. A hot object emits radiation at many wavelengths. Why did Venus seem hot only at radio wavelengths? How could the surface of Venus be kept so hot? And finally—since psychological factors may be unconsciously compelling, even in science—a Venus hotter than the hottest household oven is simply less pleasant a prospect than the Venus populated, in the long tradition from Kant to Adamski, by gracious humans of amorous or spiritual inclinations.

This problem of the origin of the Venus radio emission was a major part of my doctoral dissertation. I wrote some twenty scientific papers concerning it between 1961 and 1968, when the problem was finally considered settled. I look back on this period with pleasure. The Venus radio story is very much like a detective story where there are clues littering the pages. Some are vital to the solution; others are false clues, leading in the wrong direction. Sometimes the right answer can be deduced by bearing in mind all the relevant facts and requiring reasonable logical consistency and plausibility.

There were several things we knew about Venus. We knew

how the brightness temperature varied with radio frequency. We knew how Venus reflected back to Earth radio waves sent out by large radar telescopes. Man's first successful planetary probe—the United States' *Mariner* 2—found in 1962 that Venus was brighter at radio wavelengths at its center than at its edge.

To be matched against such observations were a variety of theories. They fell into two general categories: The hot-surface model, in which the radio emission came from the solid surface of the planet; and the cold-surface model, in which the radio emission came from somewhere else—from an ionized layer in the atmosphere of Venus, or from electrical discharges between droplets in the clouds of Venus, or from a hypothesized great belt of rapidly moving electrically charged particles surrounding Venus (like those that, in fact, surround the Earth and Jupiter). These latter models permitted the surface to be cold by placing the intense radio emission above the surface. If you wanted sailing ships on Venus, you were a cold-surface model advocate.

We systematically compared the cold-surface models with the observations and found that they all ran into serious troubles. The model in which the radio emission came from the ionosphere, for example, predicted that Venus should not reflect radio waves at all. But radar telescopes had found radio waves reflected from Venus with an efficiency of 10 or 20 percent. To circumvent such difficulties, advocates of the ionospheric model constructed very elaborate hypotheses in which there were many ionized layers with especially constructed holes in them to let radar through the ionosphere, hit the surface of Venus, and return to Earth. At the same time there could not be too many holes; otherwise, the radio emission would not be as intense as observed. These models seemed to me to be far too detailed and arbitrary in their requirements.

Just before the remarkable spacecraft observations of Venus of 1968, I submitted a paper to *Nature,* the British scientific journal, in which I summarized these conclusions and deduced that only the hot-surface model was consistent with all the evidence. I had earlier proposed a specific theory, in terms of the greenhouse effect, to explain how the surface of Venus could be at such high temperatures. But my conclusions against cold-surface models in 1968 did not depend upon the validity of the greenhouse explanation: It was just that a hot surface explained the data and a cold surface did not. Because of my interest in exobiology, I would have preferred a habitable Venus, but the facts led elsewhere. In a paper published in 1962, I had concluded from indirect evidence that the average surface temperature on Venus was about 800 degrees F and the average surface atmospheric pressure about fifty times larger than at the surface of Earth.

In 1968, an American spacecraft, *Mariner 5*, flew by Venus, and a Soviet spacecraft, *Venera 4*, entered its atmosphere. By the year 1974 there had been five Soviet instrumented capsules that entered the Venus atmosphere. The last three touched down and returned data from the planetary surface. They were the first craft of mankind to land on the surface of another planet. The average temperature on Venus turns out to be about 900 degrees F; the average pressure at the surface appears to be about ninety atmospheres. My early conclusions were approximately correct, just slightly too conservative.

It is interesting, now that we know by direct measurements the actual conditions on Venus, to read some of the criticism of the hot-surface model published in the 1960s. The year after receiving my Ph.D., I was offered, by a well-known planetary astronomer, ten-to-one odds that the surface pressure on Venus was no more than ten times that on Earth. I gladly offered my ten dollars against his hundred; to his credit, he paid off—after the Soviet landing observations were in hand.

Theory and spacecraft interact in other ways. For example, *Venera 4* radioed its last temperature/pressure point at 450 degrees F and twenty atmospheres. The Soviet scientists concluded that these were the surface conditions on Venus. But ground-based radio data had already shown that the surface temperature must be much higher. Combining radar with *Mariner 5* data, we knew that the surface of Venus was far below where the Soviet scientists concluded *Venera 4* had landed. It now appears that the designers of the first Venera spacecraft, believing the models of cold-surface theoreticians, built a relatively fragile spacecraft, which was crushed by the weight of the Venus atmosphere far above the surface—much as a submarine, not designed for great depths, will be crushed at the ocean bottoms.

At the 1968 Tokyo meeting of COSPAR, the Committee on Space Research of the International Council of Scientific Unions, I proposed that the *Venera 4* spacecraft had ceased operating some fifteen miles above the surface. My colleague, Professor A. D. Kuzmin, of the Lebedev Physical Institute, in Moscow, argued that it had landed on the surface. When I noted that the radio and radar data did not put the surface at the altitude deduced for the *Venera 4* touchdown, Dr. Kuzmin proposed that *Venera 4* had landed atop a high mountain. I argued that ground-based radar studies of Venus had shown mountains a mile high, at most, and that it was exceptionally unlikely *Venera 4* would land on the only fifteen-mile-high mountain on Venus, even if such a mountain were possible. Professor Kuzmin replied by asking me what I thought was the probability that the first German bomb to fall on Leningrad in World War II would kill the only elephant in the Leningrad zoo. I admitted that the chance was very small, indeed. He responded, triumphantly, with the information that such was indeed the fate of the Leningrad elephant.

The designers of subsequent Soviet entry probes were, despite the Leningrad zoo, cautious enough to increase the structural strength of the spacecraft in each successive mission. *Venera 7* was able to withstand pressures of 180 times that at the surface of the Earth, a quite adequate margin for the actual Venus surface conditions. It transmitted twenty minutes' worth of data from the Venus surface before being fried. *Venera 8*, in 1972, transmitted more than twice as long. The surface pressure is not at twenty atmospheres, and the spectacular Mount Kuzmin does not exist.

The principal conclusion about the scientific method that

I draw from this history is this: While theory is useful in the design of experiments, only direct experiments will convince everyone. Based only on my indirect conclusions, there would today still be many people who did not believe in a hot Venus. As a result of the *Venera* observations, everyone acknowledges a Venus of crushing pressures, stifling heat, dim illumination, and strange optical effects.

That our sister planet should be so different from Earth is a major scientific problem, and studies of Venus are of the greatest interest in understanding the earliest history of Earth. In addition, it helps to calibrate the reliability of astral projection and spirit travel of the sorts popularized by Emanuel Swedenborg, Annie Besant, and innumerable present-day imitators, none of whom caught a glimmering of the true nature of Venus.

13.
Venus Is Hell

The planet Venus floats, serene and lovely, in the sky of Earth, a bright pinpoint of yellowish-white light. Seen or photographed through a telescope, a featureless disc is discerned; a vast unbroken and enigmatic cloud layer shields the surface from our view. No human eye has seen the ground of our nearest planetary neighbor.

But we now know a great deal about Venus. From radio telescope and space-vehicle observations, we know that the surface temperature is about 900 degrees Fahrenheit. The atmospheric pressure at the surface of Venus is about ninety times that which we experience at the surface of the Earth. Since the planet's gravity is about as strong as the Earth's, there are about ninety times more molecules in the atmosphere of Venus as in the atmosphere of Earth. This dense atmosphere acts as a kind of insulating blanket, keeping the surface hot through the greenhouse effect and smoothing out temperature differences from place to place. The pole of Venus is probably not significantly colder than its equator, and on Venus it is as hot at midnight as at noon.

Forty miles above the surface is the thick cloud layer that we see from Earth. At least until recently, no one knew the composition of these clouds. I had proposed that they were constituted in part of water, a cosmically very abundant material, which could account for many but by no means all of the observed properties of the Venus clouds. But there were many other candidate materials proposed, among them, ammonium chloride, carbon suboxide, various silicates and oxides, solutions of hydrochloric acid, a hydrated ferric chloride, carbohydrates, and hydrocarbons. These last two materials were proposed by Immanuel Velikovsky in his speculative romance *Worlds in Collision* to provide manna for the Israelites during their forty years of wandering in the desert. The other candidate materials were proposed on somewhat firmer grounds. Yet each of them ran afoul of one or more of the observations.

But recently a material has been proposed that is in excellent quantitative agreement with all of the measurements. The American astronomer Andrew T. Young has shown that the clouds of Venus are very likely a concentrated solution of sulfuric acid. A 75 percent solution of H_2SO_4 precisely matches the index of refraction of the Venus clouds determined by polarimetric observations from the Earth. None of the other materials comes close. Such a solution is liquid at the temperatures and pressures at which the Venus clouds reside. Sulfuric acid has an absorption feature, determined by infrared spectroscopy, at a wavelength of 11.2 microns. Of all the materials proposed, only H_2SO_4 has such an absorption feature. The Soviet entry spacecraft of the *Venera* series have found large quantities of water vapor below the visible clouds of Venus. Ground-based observers looking for water vapor spectroscopically have found only a tiny amount of water vapor above the clouds of Venus. The two observations are in accord only if a very effective drying agent is present between these two regions. Sulfuric acid is such an agent.

In the Earth's atmosphere there are water droplets at altitude, and water vapor in the atmosphere below. Likewise on Venus: If there are sulfuric acid droplets in the high clouds, there must be gaseous sulfuric acid below, with a relatively high concentration near the surface. Astronomers in Earthbound observatories have also found unmistakable evidence of hydrochloric acid and hydrofluoric acid as gases in the upper atmosphere of Venus. They also must exist in a fair concentration–for example, the relative proportions of Los Angeles smog in Los Angeles air–in the lower atmosphere of Venus. These three acids are an extremely corrosive mixture. Any spacecraft that is to survive on the Venus surface must

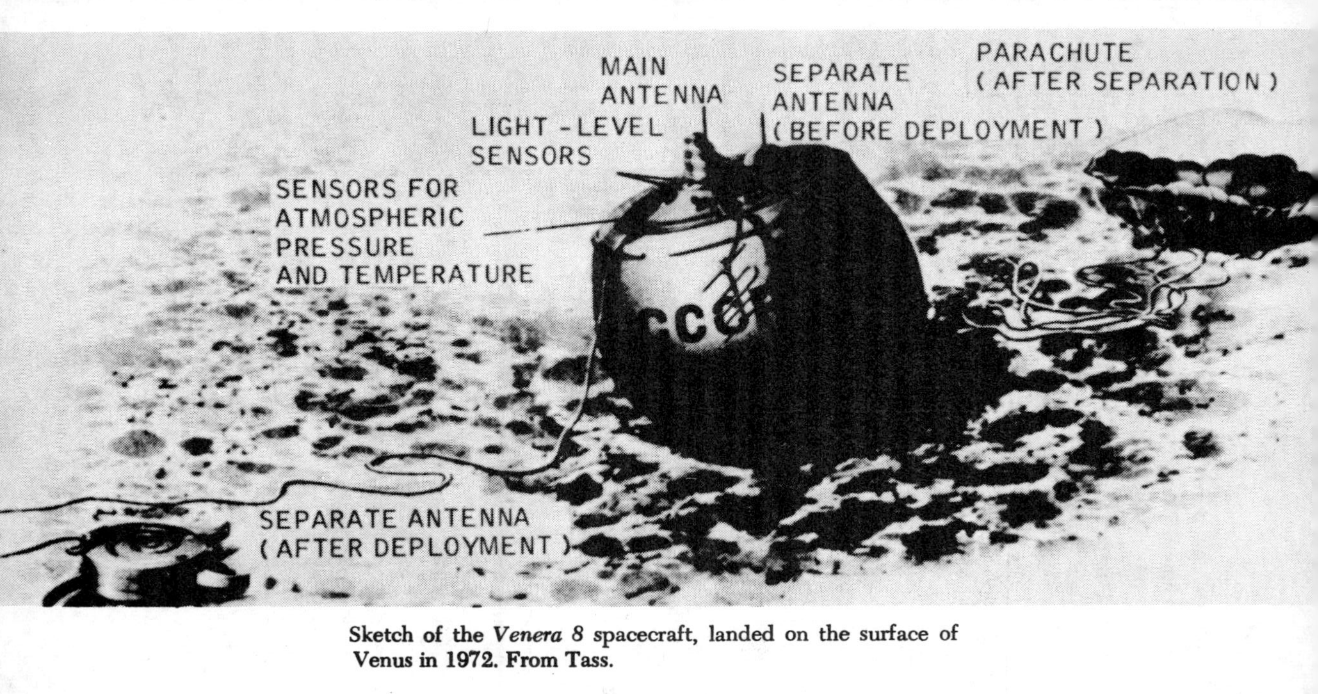

Sketch of the *Venera 8* spacecraft, landed on the surface of Venus in 1972. From Tass.

not only be bulwarked against the high pressures but protected against the corrosive atmosphere.

The Soviet Union is engaged in a very active program of unmanned exploration of Venus. We now know there is enough light for photography at midday on the Venus surface. The time will come, in not too many years, I think, when we will have our first photographs of the surface of Venus. What does the surface of Venus look like? To some extent we can already make predictions.

Because of the very dense atmosphere of Venus, there are some interesting optical effects. The most important such effect is due to Rayleigh scattering, named after the British Lord Rayleigh. When sunlight strikes the clear, dust-free atmosphere of the Earth, it is scattered. Photons strike the molecules of the Earth's atmosphere and are bounced off. Many such bounces may occur. But because the molecules of air are very much smaller than the wavelength of light, it turns out that short wavelengths are scattered or bounced away by the air molecules more efficiently than long wavelengths. Blue light is scattered much better than red light. This was a fact known to Leonardo da Vinci, who painted distant landscapes in an exquisite cerulean blue. It is why we talk of purple mountains; it is why the sky is blue. The light from the sun is scattered about in our atmosphere—some of it being scattered up and out again, but other fractions of sunlight being scattered about by the molecules of our atmosphere and then, from quite a different direction than that of the Sun, scattered back down to our eyeballs. In the absence of an atmosphere, as on the Moon, the sky is black. When we look at a sunset we are seeing the Sun through a longer path in the Earth's atmosphere than when we view it at noon. Blue light has been preferentially scattered out of this path, leaving only the red light to strike our eyes. The beauty of the sunset, the sky, and distant landscapes are all due to Rayleigh scattering.

What about Rayleigh scattering on Venus? Because the atmosphere is so much denser, Rayleigh scattering there is much more important. Were we to strip the clouds off Venus, we would still be unable to see its surface from above. Visible light of all colors would be scattered so many times in the Venus atmosphere that no image of any surface details would be discerned. In the near infrared, at wavelengths longer than the human eye is sensitive to, the surface could, however, be seen from above. But there are clouds. Radio waves penetrate the clouds and the atmosphere of Venus and the first radar maps of Venus are being developed (see page 80). In a few years, Cornell University's great Arecibo telescope in Puerto Rico will begin mapping the surface of Venus by radar with higher precision than the best ground-based optical maps of the Moon. Already, there are hints of mountain ranges and great impact basins on the surface of this enigmatic planet.

At the surface of Venus, Rayleigh scattering is also an extremely important effect. Just as we cannot see the surface in visible light from above Venus, we cannot see the sun in visible light from the surface of Venus—even if there were a break in the clouds. If there were intelligent life on Venus, astronomy would be very slow to develop; and radio astronomy would emerge first. The *Venera 8* spacecraft found that sunlight does reach the surface of Venus during the day, but it is so attenuated by passage through the clouds and atmosphere that, even at midday, it is no brighter on Venus than at twilight on the Earth. The sunlight would be a hazy and diffuse patch of deep ruby-red light, whose rising and setting could only indistinctly be determined.

If you were standing in some protective suit on the surface of Venus and put on violet sunglasses, you would see no farther than a few dozen feet. The Rayleigh scattering in blue light is so strong on Venus that the visibility in the violet is small. But because long wavelength light is scattered less than blue, at the extreme red end of the visible spectrum—with red sunglasses on—you could see perhaps a thousand feet. At the surface of Venus everything would be suffused in a deep red gloom. We would have a perception of color, but only for objects very close to us. Our surroundings would be an indistinct roseate blur.

Venus thus seems to be a place quite different from the Earth, and alarmingly unappealing: Broiling temperatures, crushing pressures, noxious and corrosive gases, sulfurous smells, and a landscape immersed in a ruddy gloom.

Curiously enough, there is a place astonishingly like this in the superstition, folklore and legends of men. We call it Hell. In the older belief—that of the Greeks, for example—it was the place where all human souls journeyed after death. In Christian times it has been thought of as the post-mortem destination only of one of two categories of moral persuasion. But there is little doubt that the average person's view of Hell—sizzling, choking, sulfurous, and red—is a dead ringer for the surface of Venus.

Although terrestrial biological molecules would fall to pieces rapidly on Venus, there are organic molecules—for example, some with a complex ring structure—that would be quite stable under the conditions of Venus. It is difficult to exclude life there, but we can certainly say it would be quite different from what we are familiar with. Any organism that lives there would be wise to have leathery skin. Because of the high atmospheric pressures, it would even make sense to have little stubby wings, which could carry their possessors about without exceptionally strenuous flapping. A devil is a very good model—except for his mannish and goaty aspects—for an inhabitant of Venus. Milton and Isaiah called Lucifer "Son of the Morning," the morning star. For thousands of years Venus and Hell have been identified.

This is all a very curious coincidence, but I cannot bring myself to think that it is anything more than that. The chief point is that in all the legends one gets to Hell by going down, not up. The classical world of Greece and Rome and the ancient Near East were peppered with active volcanoes. Such volcanic terrains, like in contemporary Iceland and Hawaii, are bleak, desolate, and eerily beautiful landscapes. Sulfurous gases emanate from volcanic vents; lava fountains and flows suffuse the surroundings in red. It is very hot: You singe your eyebrows if you get too close to the lava in a collapsed lava tube. And all this heat, redness, and smell come from down below. It was not very difficult for our ancestors to imagine that volcanic terrains were apertures to a quite different igneous world called Hell.

The inside of the Earth and the outside of Venus are alike but not identical. They are both unpleasant for humans. But they are both of extreme scientific interest—worth at least an extended visitation, if not a homesteading. Dante knew about that.

Assorted foreign military attachés assembled to view the *Apollo 15* launch. Photograph by the author.

14. Science and "Intelligence"

I spent my first two postdoctoral years at the University of California, Berkeley, where, among other things, I was concerned with searching for life elsewhere and with the sterilization of space vehicles intended for places like Mars—we wished not to contaminate the Martian environment with microbes from Earth.

One bright spring day I received a phone call from an Air Force general whom I had met at several scientific meetings. He had been working chiefly on aviation medicine; I will call him here Bart Doppelganger. General Doppelganger informed me that he was in Los Angeles with three Soviet scientists, one of whom was in charge of the Soviet effort for constructing instruments to search for extraterrestrial life. His name was Alexander Alexandrovitch Imshenetsky (there is no reason to change his name; unlike some others in this narrative he has nothing to be ashamed of). It was Imshenetsky's first visit to the United States. Yes, I would certainly be interested in meeting him. When? The answer was "immediately." So I drove to San Francisco airport, flew to Los Angeles, and took a taxi to an address given to me by our General Doppelganger.

It was the home of a professionally well-known UCLA brain physiologist. In the living room, upon my arrival, were the physiologist, other aviation medicine experts from UCLA, General Doppelganger, three Soviet scientists (two in aviation medicine and Academician Imshenetsky), and a translator. I will call the translator Igor Rogovin; he was an American employee of the Library of Congress assigned to do translation for the three Russians on their visit to the United States. The only thing that struck me as somewhat peculiar was that the English of all three Russians was quite good. So why a translator?

Everyone was jolly, pleasantries were exchanged, drinks made the rounds—and Igor Rogovin also made the rounds. There was no conversation from which he was absent for more than a few minutes. He was very busy, like a manic bee obsessively flitting from flower to flower.

After a while, the plan was for all of us to drive to Los Angeles International Airport, where the Russians were later to catch a plane. Before their flight, we were all to have dinner. There were more of us than could fit into one car, and Rogovin could not easily ride in both cars at once. Imshenetsky, some others, and I rode in one car, and Rogovin and the others in a second. During the twenty- or thirty-minute drive, Imshenetsky and I had a fruitful exchange of views on methods of life detection and space-vehicle sterilization technology. It was the first such contact I had had with a Soviet scientist.

We arrived at the airport, bags were checked, and the Soviets excused themselves to go to the men's room. Waiting outside, I found myself alone with Igor Rogovin—who immediately said to me out of the corner of his mouth, and in a style of vocalization that went out with James Cagney and Humphrey Bogart, "Hey, kid, what'd ya find out?"

Being unwise in the ways of the world, and pleased with the information Imshenetsky and I had exchanged, I rapidly summarized what I had learned.

"Pretty good, kid. Who do you work for?"

"The University of California at Berkeley," I replied brightly.

"No, no, kid, not the cover."

Igor Rogovin's occupation, if not his identity, gradually dawned on me. With a rising fury I explained to him that it was possible to have a conversation with a Soviet scientist that was intended for the benefit of science rather than for the benefit of American military intelligence services. Before

Rogovin could reply, our friendly Soviet guests re-emerged, and we all went off to dinner.

Although seated again next to Imshenetsky, I found myself unable to talk to him on any subject remotely approaching science. As I recall, our primary topics of conversation were American films and Soviet poets. After a number of drinks, Alexander Alexandrovitch Imshenetsky offered the opinions that William Shakespeare was the leading Russian poet and American cowboy films were excessively violent. Several hours easily could be spent discussing these two propositions. As the Russians left to fly home, I returned to Berkeley.

The next morning, I turned to the white pages of the San Francisco telephone directory and, under "United States Government," found a section marked "Central Intelligence Agency." On dialing the number indicated I encountered a cheerful voice that said something like "Yukon 4-2143."

"Hello, Central Intelligence Agency?" I said.

"What can we do for you, sir?"

"I have a complaint to file."

"One moment, sir, I will give you our Complaint Department." This was shortly after the Bay of Pigs, and I guess they had been getting a lot of complaints.

Upon reaching—yes—the Complaint Department, I rapidly launched into a synopsis of my encounter with Mr. Rogovin, but was quickly silenced with an injunction that this was not the sort of thing one talked about over the telephone. I suppose his phone was tapped. We made an appointment for later that afternoon in *my* office.

Sure enough, at the prescribed time, two neatly dressed business-suited young men arrived, bearing plastic identification cards with the signature of John McCone, who had recently been appointed Director of the Central Intelligence Agency. After expressing my general annoyance by an exceptionally long scrutiny of their IDs, I launched into my story. In their faces was mounting concern. At the conclusion of my narrative, they explained to me that Rogovin's behavior was very unlikely to be the behavior of any employee of "the agency." They were very concerned about such a story, particularly after the "bad press" they had been getting about the Bay of Pigs. They would do everything to track the story down if I would not embarrass the agency by further disseminating the story. I agreed to keep quiet, for a while, and they departed.

About a week later I received a phone call: "Dr. Sagan, this is Mr. Smith, who was in your office last week. You recall the matter about which we talked?" I gathered that his phone was still tapped.

"We have been able to establish that the party in question—you know who I mean?—does not work for our organization under that name. We are, of course, pursuing other names and will get back to you as soon as we can."

It had taken them a week to scrutinize the personnel roster of the Central Intelligence Agency. The roster must either be very long or very secret.

Several days later, in similarly veiled language, I was called and told—in a voice that seemed to express considerable concern—that Igor Rogovin did not work for the CIA under any name whatever, and that they were, under the circumstances, naturally curious to know for whom he did work.

After another week had passed, the CIA made an appointment to speak with me once again in my office. The same two gentlemen arrived, again bearing the same two plastic cards signed by John McCone. They informed me that, after exhaustive investigations, they had discovered that Igor Rogovin, while nominally working for the Library of Congress, was, in fact, an employee of Air Force Intelligence. They again assured me that no representative of their agency would ever behave in so uncouth a manner, and departed. What stood out most clearly was that it had taken about two weeks for the Central Intelligence Agency to determine the employment of a member of a fellow U.S. intelligence organization.

This story has a coda. A year or two later the international space organization, COSPAR, was meeting in Florence, Italy. In one of the splendid side benefits of such meetings, the Uffizi Gallery was opened one evening especially for the members of the COSPAR delegations from various nations. As chance would have it, I was with Alexander Alexandrovitch Imshenetsky as we entered an enormous and apparently empty Botticelli-inundated gallery. And there, at the other end of the gallery, a human figure could dimly be perceived. I felt Imshenetsky stiffen. Peering intently, I could now make out the visage of Igor Rogovin, his face disguised by a beard. He was, no doubt, traveling incognito. Imshenetsky leaned over to me and whispered, "Isn't that the fellow who was with us in Los Angeles?" When I nodded assent, Imshenetsky murmured, "Very stupid fellow."

That Rogovin was working for Air Force Intelligence is, in retrospect, not so surprising, since I had been invited to Los Angeles by General Doppelganger. That Soviet plans for the search for life elsewhere or for the sterilization of spacecraft could be considered of interest to Air Force Intelligence is perhaps more surprising. That American intelligence agencies would attempt to use comparatively innocent young scientists (I was twenty-seven and politically unsophisticated) to carry out such a purpose is appalling. At least it appalls me.

There are many other such stories involving both American and Soviet intelligence organizations. The general effect of such incidents is to detract from the credibility of legitimate scientific exchanges among scientists of different countries. Such exchanges are particularly necessary in an age that hangs a thread away from nuclear destruction, and in which scientists have access to at least half an ear of the politicians in power. The fact that such intelligence activities are practiced in an entirely regular and invariable manner on the Soviet side does not, in my view, weaken this argument. The intrusion of "intelligence" into international scientific exchanges of this sort is, whatever else it is, just not intelligent.

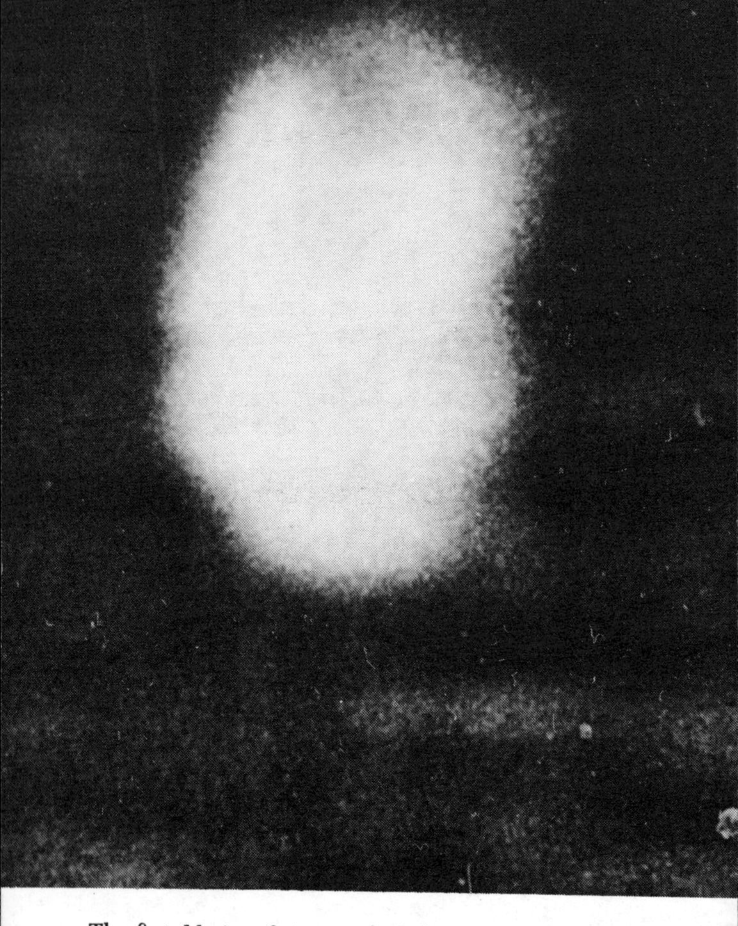

The first *Mariner* 9 image of Phobos, unprocessed by computer. (See also photos on pages 102 and 103.)

15.
The Moons of Barsoom

In my boyhood I was lucky enough to come upon a set of turgidly written novels with names like *Thuvia, Maid of Mars, The Chessman of Mars, The Princess of Mars, The Warlords of Mars,* and so on. They were, needless to say, about Mars. But they were not about *our* Mars—the Mars revealed by *Mariner* 9.

At least I don't think our Mars is like the Mars of these novels by Edgar Rice Burroughs, the inventor of Tarzan. His Mars was Percival Lowell's Mars—a planet of ancient sea bottoms, working canals and pumping stations, six-legged beasts of burden, and men (some headless) of all colors, including green. They had names like Tars Tarkas. Possibly the most remarkable hypothesis proposed by Burroughs in these novels was that human beings and inhabitants of Mars could produce live offspring—a biologically impossible proposition if the Martians and we are imagined as having separate biological origins. Burroughs wrote decorously of the interfertility of a Virginian miraculously transported to Mars and Dejah Thoris, the princess of a kingdom with the improbable name of Helium. I have little doubt that the precedent of a kingdom called Helium led directly to the planet called Krypton, home of the comic-book hero Superman. There is here a rich vein of untapped literary ore. The future may hold planets, stars, or even entire galaxies named Neon, Argon, Xenon, and Radon—the remaining noble gases.

But the name invented by Burroughs that has haunted me across the years is the name he imagined the Martians gave

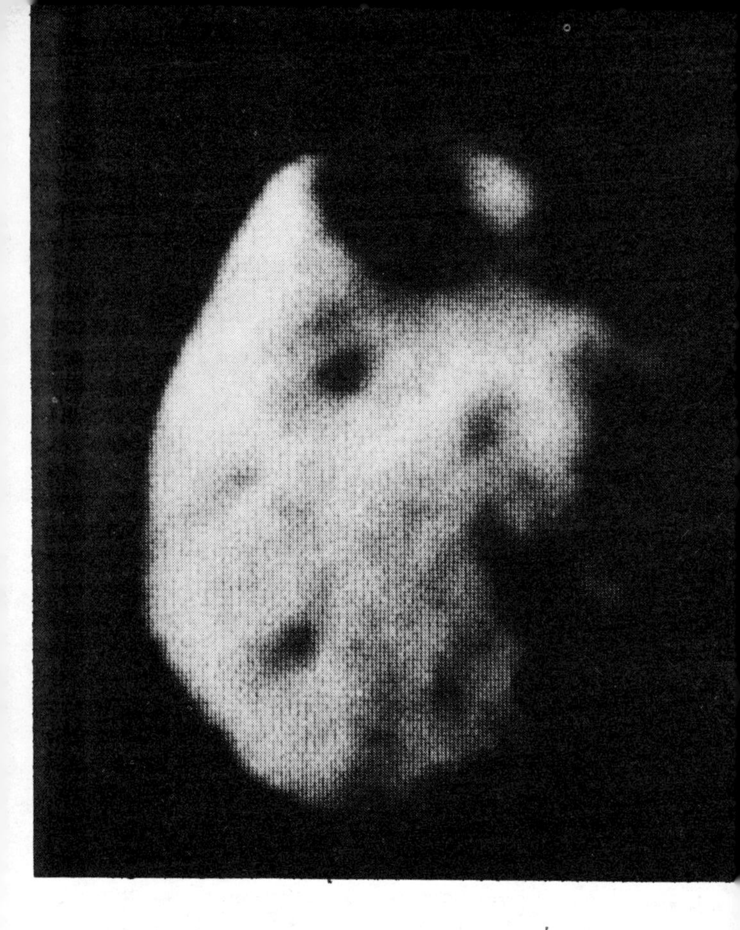

The same photograph as on page 100, contrast-enhanced by computer, and showing cratering.

to Mars: Barsoom. And it was one phrase of his more than any other that captured my imagination: "The hurtling moons of Barsoom."

For Mars is indeed a world with two moons—a situation that would appear to the inhabitants of Mars as entirely natural as our one Moon does to us. We know how our solitary satellite looks to the naked eye from the surface of Earth. But

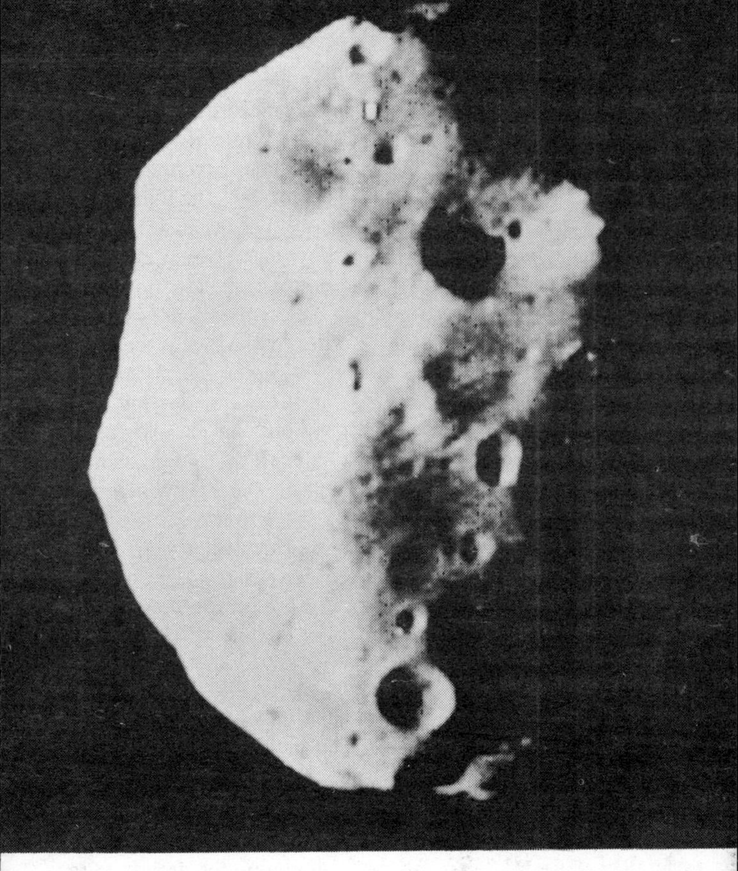

A later, much more detailed image of Phobos showing its true nature.

what do the moons of Barsoom look like from the surface of Mars? This question, which intermittently plagued my boyhood, was not to be answered until 1971 and *Mariner 9*.

The moons of Mars were invented by Johannes Kepler, the discoverer of the laws of planetary motion and no intellectual lightweight. But he lived in the sixteenth century, in a different intellectual climate from the present. He cast horoscopes for a living; astronomy was his passion more than his occupation. His mother was tried as a witch. When Kepler learned of Galileo's discovery, with one of the first astronomical telescopes, of the four large moons of Jupiter, he immediately concluded that Mars had two moons. Why? Because Mars was at an intermediate distance from the Sun, between Earth and Jupiter. It must obviously have an intermediate number of moons. The observations seemed to show Venus with no moons, Earth with one, and Jupiter with four (the actual number, we now know, is twelve). Kepler could have deduced either two or three moons for Mars. But bearing a lifelong passion for geometrical progressions, he chose two. The argument is, of course, fallacious. Saturn's ten moons, Uranus' five, and Neptune's two in no way fit his scheme, which is not scientific but aesthetic.

But Kepler's prestige was immense, particularly after Kepler's laws of planetary motion were derived from gravitational theory by Isaac Newton; and so, literary allusions to the two moons of Mars fluttered down the centuries. In Voltaire's longish short story "Micromegas" a denizen of the star Sirius notes casually, while touring our Solar System, that Mars had two moons. There is a more famous reference to two Martian moons in Jonathan Swift's satire of 1726, *Gulliver's Travels*—not the part about the very small people, or the part about the very large people, or the part about the intelligent horses, but a less widely read part—the part about the floating aerial island of Laputa. The episode is undoubtedly a tightly reasoned critique of English-Spanish relations in Swift's time, because *la puta* is Spanish for prostitute. The political metaphors are obscure, at least to me. At any rate, Swift announces casually that astronomers on Laputa have discovered two swiftly moving moons of Mars, and have provided information on their distances from Mars and their periods of revolution about Mars—information that is incorrect, but that is not a bad guess. There is an entire genre of writing on how it was that Swift knew about the moons of Mars, including the suggestion that he was a Martian. Internal evidence suggests that

Swift was no Martian, and the two moons can almost certainly be traced directly back to Kepler's speculations.

The actual discovery of the moons of Mars was made from outside Washington, D.C., in 1877. The U. S. Naval Observatory had just completed a large refracting telescope. The observatory's astronomer, Asaph Hall, sought to learn whether the moons of Mars, fabled in song and story, actually existed. His first few nights were unsuccessful, and, in a despondent mood, he announced to his wife his intention of quitting the search. Mrs. Hall would have none of this, and urged her reluctant husband back to a few more nights at the telescope—whence he emerged with two Martian satellites in hand. For a brief period he thought he had found three satellites, because the inner one moved so rapidly that on one night he saw it on one side of Mars and on the next night on the other. Hall named the moons Phobos and Deimos, after the horses that pulled the chariot of the god of war in Greek mythology; they have the cheerful denotations of fear and terror, respectively. (The appropriate adjectives pose some problems: Do we talk of Phobic orbits and Deimonic nights?) If another moon of Mars is someday found, I hope we will give it a less ferocious and more optimistic name—like "Pax."

I also hope that when the features on Phobos and Deimos are eventually named by the International Astronomical Union, one will be named after Mrs. Hall. But since another feature will surely be named after Asaph Hall, we have a problem—two craters named Hall would be confusing. In an astronomy talk I gave at Harvard University, I commented wistfully that the problem would be solved if only we knew Mrs. Hall's maiden name. My friend Owen Gingerich, Professor of the History of Science at Harvard, instantly leaped to his feet with the words "Angelina Stickney" tripping off his lips. So when the time comes, I hope there will be a "Stickney" on one of the moons of Barsoom.

The subsequent study of Phobos and Deimos between 1877 and 1971 has a curious history. The moons of Mars are so tiny that they appear, even to the largest Earthbound telescopes, as dim points of light. They are too faint for the pre-1877 telescopes to have seen them at all. Their orbits can be calculated by noting their positions at various times. In 1944 at the U. S. Naval Observatory (where an understandably proprietary interest in Phobos and Deimos must have developed), B. P. Sharpless collected all the observations available in his day to determine the orbits to the best possible precision. He found—no doubt to his surprise—that the orbit of Phobos appeared to be decaying, what astronomers call a secular acceleration. Over longish periods of time the satellite seemed to be approaching more and more closely to Mars and moving more and more rapidly. This phenomenon is quite familiar to us today. The orbits of artificial satellites are decaying all the time in the Earth's atmosphere. They are initially slowed by collisions with the diffuse upper atmosphere of the Earth, but by Kepler's laws the net result is a more rapid motion.

Sharpless' conclusion of a secular acceleration for Phobos remained an unexplained and almost unexamined curiosity until it was considered around 1960 by the Soviet astrophysicist I. S. Shklovskii. Shklovskii considered a wide range of alternative hypotheses for the secular acceleration, among them the influence of the Sun, the influence of a hypothetical magnetic field on Mars, and the tidal influence of the gravity of Mars. He found that none of these came close to working. He then reconsidered the possibility of atmospheric drag. The exact size of the Martian satellites was known poorly and indirectly in those days before spacecraft investigation of Mars, but it was known that Phobos was roughly ten miles in diameter. The altitude of Phobos above the surface of Mars was also known. Shklovskii and others before him found that the density of the atmosphere was far too low to produce the drag that Sharpless had deduced. It was at this point that Shklovskii made a brilliant and daring guess.

All the calculations showing atmospheric drag to be ineffectual had assumed that Phobos was an object of ordinary density. But what if its density were very low? Despite its enormous size, its mass would then be quite small, and its orbit could be appropriately affected by the thin upper atmosphere of Mars.

Shklovskii calculated the required density of Phobos, and found a value about one one-thousandth the density of water. There is no natural object or substance with such low density; balsa wood, for example, has about half the density of water. With such a low density, there was only one conclusion possible: Phobos had to be hollow. A vast hollow object ten miles across could not have arisen by natural processes. Shklovskii, therefore, concluded that it was produced by an advanced Martian civilization. Indeed, an artificial satellite ten miles across requires a technology far in advance of our own; it would also be far in advance of the technology imagined on Barsoom by Burroughs, which was a kind of sword and small spaceship technology.

Since there were no signs of such an advanced civilization on Mars today, Shklovskii concluded that Phobos—and possibly Deimos—had been launched in the distant past by a now extinct Martian civilization. (The interested reader may find more details of this remarkable argument of Shklovskii's in the book *Intelligent Life in the Universe,* jointly authored by Shklovskii and myself. [San Francisco, Holden-Day, 1966; New York, Delta Books, 1967]). Subsequent to Shklovskii's first work on the subject, the motions of the moons of Mars were re-examined in England by G. A. Wilkins, who found that possibly there was no secular acceleration. But he could not be sure.

Shklovskii's extraordinary suggestion that the moons of Mars might be artificial is one of three hypotheses on their origin. The other two—certainly interesting in their own right, but naturally paling in comparison with the Shklovskii hypothesis—are (1) that the moons are captured asteroids, or (2) that they are debris left over from the origin of Mars itself.

Asteroids are hunks of rock and metal that go around the Sun between the orbits of Mars and Jupiter. There are unlikely, but theoretically possible, scenarios in which the gravitation of Mars can capture a close-passing asteroid.

In the Martian-debris hypothesis, it is imagined that pieces of rock of various sizes fell together to form Mars; that the last generation of such infalling pieces produced the large old impact craters on Mars (see page 131); and that Phobos and Deimos are by chance the only remnants still extant of the early catastrophic history of Mars.

It is clear that establishing any one of these three hypotheses on the origin of the moons of Mars would be a major scientific achievement.

The Mariner Mars mission of 1971, which I had the pleasure to work on, was originally to have involved two spacecraft, *Mariner 8* and *Mariner 9*. They were to be placed in different orbits for different purposes in the study of Mars itself. After these orbits were finally agreed upon, I noticed that they were not all that far from the orbits of Phobos and Deimos. It also seemed to me that television and other close-up observations of Phobos and Deimos by the Mariner spacecraft might permit us to determine something of their origin and nature.

I therefore approached officials of NASA, which organized and ran the mission, for permission to program observations of Phobos and Deimos. While the mission controllers at Jet Propulsion Laboratory, the actual operating organization, were not unsympathetic to this idea, some officials at NASA headquarters were against it. There was a mission plan, written in a large book, which stated what *Mariners 8* and *9* were about. Nowhere in the mission plan were Phobos and Deimos mentioned. Ergo, I could not look at Phobos and Deimos.

I pointed out that my proposal required only moving the scan platforms on the spacecraft so that the cameras could observe the Martian satellites. The response was negative again. A short time later, I advanced the argument that if Phobos and Deimos were indeed captured asteroids, examining them from *Mariner 9* was the equivalent of a free mission to the asteroid belt: The proposed scan platform maneuver would save NASA two hundred million dollars or so. This argument was judged, at least in some circles, to be more compelling. After about a year of my lobbying, a planning group on satellite astronomy was set up, and tentative plans were made for examining Phobos and Deimos. The satellite astronomy working group was, at my suggestion, chaired by Dr. James Pollack, a former student of mine; but it was a sign of NASA reluctance that the group was formed only after the launch of *Mariner 9*, and only about two months before its arrival at Mars. (*Mariner 8* had, meanwhile, failed.)

When *Mariner 9* arrived at Mars, we found a planet almost entirely obscured by dust. Since there was little to look at on Mars, a great and previously undetectable enthusiasm for examining Phobos and Deimos dramatically materialized. The first step was to take wide-angle photographs from a distance in order to establish with some precision the orbits and locations of the moons. This task was accomplished in a preliminary way about two weeks after injection of the spacecraft into Martian orbit. *Mariner 9* has an orbital period of about twelve hours, so that it made close to two revolutions around Mars per day.

The television pictures from *Mariner 9* were radioed from Mars to Earth in much the same way that a newsprint wirephoto is transmitted on Earth. The picture is divided into a large number of small dots (for *Mariner 9*, several hundred thousand dots), each dot with its own brightness, or shade of gray, running from black to white. After the picture is taken by the spacecraft and recorded there on magnetic tape, it is played back to Earth, dot by dot. The communication says, in effect: Dot number 3277, gray level 65; dot number 3278, gray level 62, and so on. The picture is reassembled by computer on the Earth—essentially by following the dots.

The first moderately close-up photograph of Phobos was obtained on revolution 31. Page 100 shows a Polaroid photo of the video-monitor image of Phobos on revolution 31, received on November 30, 1971. The image is much too indistinct to make any conclusions whatever.

Late that same night, Dr. Joseph Veverka, of Cornell, another former student of mine, and I worked into the small hours of the morning at the Image Processing Laboratory of JPL to bring out—by computer contrast-enhancement techniques—all of the detail present in the image. The result is shown on page 102. The shape is irregular. Are those blotches craters?

Our computer-enhanced photograph was constructed on the computer's video monitor, line by line, from top to bottom. As the apparent large crater at the top gradually emerged, we saw a single bright spot at its center; for just a moment, I had the sense that we were seeing a star through an enormous hole in Phobos—or, even more chilling, that we were seeing an artificial light. But when we requested the computer to remove all single-bit errors, the bright spot went away.

On revolution 34, *Mariner 9* and Phobos came within less than four thousand miles of each other, one of the closest approaches in the entire mission. Late on the night of the receipt of that picture, Veverka and I were again computer-enhancing. Our results were like those seen on the accompanying page 103. I am not sure what an artificial satellite ten miles across looks like, but this does not seem to be it. Phobos looks not so much like an artificial satellite as a diseased potato. It is, in fact, very heavily cratered. For it to have accumulated so many craters in that part of the Solar System, it must be very old, probably billions of years old.

Phobos appears to be an entirely natural fragment of a larger rock severely battered by repeated collisions; holes have been dug, pieces have been chipped off. It looks a little like the hand axes, chipped along natural fracture planes, made by our Pleistocene ancestors. There is no sign of technology on it. Phobos is not an artificial satellite. When pictures of Deimos were computer contrast-enhanced, the same conclusion applied to it.

Phobos and Deimos are the first satellites of another planet to have been photographed close-up. They were also observed by the ultraviolet spectrometer and the infrared radiometer on board *Mariner 9*. We have been able to determine their sizes and shapes and something of their color. They are extremely dark objects—darker than the darkest material that is likely to be in the room in which you are sitting right now.

Indeed, they are among the very darkest objects in the Solar System. Since there are so few objects this dark anywhere, we hope to be able to conclude something about their composition. They are both covered by at least thin layers of finely pulverized material. They provide important clues to collisional processes in the early Solar System. I believe we are looking at the end-product of a kind of collisional natural selection, in which fragments have been broken off from a larger parent body, and we are seeing only the two pieces, Phobos and Deimos, that remain. The moons of Mars are also important collision calibrators for Mars. Phobos, Deimos, and Mars have very likely been together in the same part of the Solar System for a very long period of time. The number of craters of a given size on Mars is much less, in general, than on Phobos and Deimos, providing important information on erosional processes that exist on Mars and that do not exist on airless and waterless Phobos and Deimos.

Because we now have the first good information on the size and shape of these objects, and because we now have good reason to think that they have typical densities of ordinary rock, we can calculate something about what it would be like to stand on, let's say, Phobos. First of all, Mars, less than six thousand miles away, would fill about half the sky of Phobos. Marsrise would be a spectacular event. Eventual construction of an observatory on Phobos to examine Mars might not be such a bad idea. We know from *Mariner 9* that both Phobos and Deimos are rotating as our Moon does, always keeping the same face to their planet. When Phobos is above the day hemisphere of Mars, the reddish light of Mars would be enough to read by at night on Phobos.

Because of their small sizes, Phobos and Deimos have very low gravitational accelerations. Their gravities do not pull very hard. The pull on Phobos is only about one one-thousandth of that on Earth. If you can perform a standing high jump of two or three feet on Earth, you could perform a standing high jump of half a mile on Phobos. It would not take many such jumps to circumnavigate Phobos. They would be graceful, slow, arcing leaps, taking many minutes to reach the high point of the self-propelled trajectory and then to return gently to the ground.

Even more interesting would be a game like baseball on Phobos. The velocity necessary to launch an object into orbit about Phobos is only about twenty miles per hour. An amateur baseball pitcher could easily launch a baseball into orbit around Phobos. The escape velocity from Phobos is only about thirty miles per hour, a speed easily reached by professional baseball pitchers. A baseball that had escaped from Phobos would still be in orbit about Mars—a man-launched moonlet. If Phobos were perfectly spherical, a lonely astronaut with an interest in baseball could invent a curious but somewhat sluggish version of this already rather sluggish game. First, as pitcher, he could throw the ball sidearm—at the horizon at between twenty and thirty miles per hour. He could then go home for lunch, because it will take about two hours for the baseball to circumnavigate Phobos. After lunch, he can pick up a bat, face the other direction and await his pitch of two hours earlier. Apart from the fact that good pitchers are seldom good hitters, hitting this pitch would be pretty easy: About fifteen seconds elapse from the appearance of the baseball at the horizon to its arrival in the vicinity of our astronaut. If he swings and misses—or, more likely, if the ball is wide of the plate—he can then go home for a two-hour nap, returning with his catcher's mitt to catch the ball. Alternatively, if he succeeds in hitting a fly ball at a velocity somewhere between twenty and thirty miles per hour, he can go home and take his nap, returning this time with a fielder's mitt, awaiting the return of the ball from the opposite horizon two hours later. Because Phobos is gravitationally lumpy, the game would be more difficult than I have indicated. Since daylight on Phobos lasts only about four hours, lights would have to be erected, or the game modified so that all pitching, hitting, and catching events happen on the day side.

These sports possibilities may, one day a century or two hence, provide a tourist industry for Phobos and Deimos. But baseball on Phobos is no more an argument for going there than, to take a random example, golf is for going to the Moon. The scientific interest in the moons of Mars—whether captured asteroids or debris from the formation of the planet—is, however, immense. Sooner or later, certainly on a time scale of centuries, there will be instruments—and then men—on the surface of Phobos looking up with awe at an immense red planet that fills the sky from zenith to horizon.

And what about the opposite view? What do the moons of Barsoom look like from the surface of Mars? Because Phobos is so close to Mars, it would be seen as a clearly discernible disc, even though it is intrinsically such a tiny object. In fact, Phobos would appear as about half the apparent size of our Moon seen from the surface of Earth. We have found from *Mariner* 9 that only one side of Phobos is visible from Mars, just as only one side of our Moon is visible from Earth. That face of Phobos is, more or less, the face on page 102. Until *Mariner* 9, no one—except Martians, if such there be—ever knew that face.

Because Phobos is so close to Mars, Kepler's laws constrain it to move comparatively rapidly about the planet. It makes approximately 2½ revolutions about Mars in 24 hours. Deimos, on the other hand, takes 30 hours 18 minutes to revolve in its orbit once about Mars. Both moons revolve in their orbits in the same direction or sense as Mars rotates on its axis. Thus, Deimos rises in the east and sets in the west as—from terrestrial chauvinism—we believe a well-behaved satellite should. But Phobos makes it once around its orbit in less time than it takes for Mars to rotate. Accordingly, Phobos rises in the west and sets in the east, taking about 5½ hours to transit from horizon to horizon. This is not exactly "hurtling"—the motion would not be easily perceptible against the field of stars in a minute's watching—but it's not plodding, either. There will be some nights at the equator on Mars when Phobos sets in the east at sunset and then rises in the west well before dawn.

Phobos is so close to the equatorial plane of Mars that it is entirely invisible from the polar regions of the planet. If we were to imagine intelligent beings developing on Mars, astronomy might very well be the province of only the equatorial, and not the high-latitude, societies. I am not sure whether Helium was an equatorial kingdom.

Freud says somewhere that the only happy men are those whose boyhood dreams are realized. I cannot say that it has made my life carefree. But I will never forget those early-morning hours in a chilly California November when Joe Veverka, a JPL technician, and I were the first human beings ever to see the face of Phobos.

The State of California was kind enough to give me an automobile license plate marked "PHOBOS." My car is not particularly sluggish, but it cannot circumnavigate our planet twice a day, either. The license plate pleases me. I would have preferred "BARSOOM," but there is a strictly enforced limit of six letters per license plate.

16.
The Mountains of Mars

I. OBSERVATIONS FROM EARTH

The mountains of the Earth are the product of ages of geological catastrophes. The major folded mountain ranges are thought to be produced by the collision of enormous continental blocks during continental drift. The motion of continents toward and away from each other, at a rate of about an inch a year, seems terribly slow to us. But since the Earth is billions of years old, there has been plenty of time for continents to bang around all over our planet.

Lesser mountains were produced by volcanic events. Hot molten rock, called lava, upwells through tubes in the upper layers of the Earth—tubes of structural weakness through which the underlying pressure is relieved—and produces large surface piles of cooling volcanic slag. The resulting hole in the top of the volcanic mountain—the geologists call it a summit caldera—is the channel through which successive episodes of lava-upwelling occur. In the summit caldera of an active volcano, as, for example, in Hawaii, we can actually see molten lava. These individual volcanic mountains and mountain ranges, which are not really separate entities, are signs of a geologically vigorous and dynamic Earth.

What about Mars? It is a smaller planet than Earth; its central pressures and temperatures are less; it has a lower average density than Earth. These circumstances combine to suggest that Mars should be geologically less active than Earth, perhaps like the Moon. But even on the Moon, a much smaller object than Mars, with even lower anticipated interior temperatures than Mars, recent signs of volcanic activity have been uncovered by the Apollo missions. We do not even today understand the connection between the size and structure of a planet and the presence of volcanism and mountains, although we do know that there are no significant folded mountain ranges on the Moon.

Our present ignorance on this subject is exceeded by the ignorance of the early planetary astronomers, less than a century ago, as they peered through small telescopes and tried to guess what distant Mars was like. One of the earliest astronomers to commit himself on the question of mountains on Mars was Percival Lowell. Lowell believed (see Chapter 18) that he had found evidence of an extensive network of straight lines, crisscrossing the Martian surface with remarkable regularity and straightness, and that could only have been produced by a race of intelligent beings on that planet. He believed that these "canals" were truly canals carrying water. We now know that the problem was not so much with his logic as with his observations; none of the Mariner or other recent quantitative observations of Mars have shown any sign of the Lowellian canals.

In the 1890s Lowell argued that Mars must have no mountains, because mountains would be a severe impediment to the construction of a comprehensive network of canals. But surely a race that could construct a planetwide network of canals should be able now and again to mow down an awkwardly placed mountain.

Nevertheless, Lowell was among the very first astronomers to apply an actual observational test to the question of mountains on Mars. He looked beyond the terminator. The terminator is the line—sharp or fuzzy, depending on the absence or presence of a planetary atmosphere—that separates the day from the night side of a planet. The terminator moves around the planet once a day—the local planetary day. But if there are mountains just on the night side of the terminator, the mountains will receive the rays of the setting Sun when their adjacent valleys are in darkness. Galileo first used this technique to discover what he called the mountains of the Moon—although the lunar mountains are mainly enormous pieces of rubble that fell out of the sky in the final phases of the

Topographical map, based on radar studies, of elevations at midlatitudes on Mars. Laboratory for Planetary Studies, Cornell University.

formation of the Moon, rather than mountains of the terrestrial type, produced by a geologically active interior.

Lowell and his collaborators found cases of bright projections beyond the Martian terminator, illuminated by the rays of the setting Sun. But when they calculated their altitudes—an easy task for anyone grounded in high school geometry—the mountains were found to be many tens of miles high. Such elevations on Mars seemed to him absurd because of his canal argument. Moreover, the next day—the day on Mars is almost exactly the same length as a day on Earth—when the feature was seen again, its position had changed. This behavior is quite uncharacteristic of mountains of whatever origin, and Lowell correctly concluded that he had been seeing dust storms, in which fine particles from the Martian surface had been carried some tens of miles into the Martian atmosphere.

Such dust storms are also observed when we examine through the telescope the day side of Mars. We sometimes see that the characteristic configuration of bright and dark markings on the planet is temporarily obscured. There is an intrusion of bright-area material into the dark area, followed by a reappearance of the former configuration. These changes were interpreted in Lowell's time as dust storms arising in the bright areas and obscuring the adjacent dark areas. The present interpretation, based on the full range of *Mariner 9* close-up observations, confirms this view (see Chapter 19).

Lowell and his contemporaries called the bright areas "deserts," and this, too, seems to be an appropriate name. The Lowellians concerned themselves with the problem of whether bright areas tended to be higher or lower than dark areas, even though the elevation difference was expected to be extremely small. A dark area seen at the illuminated limb, or edge, of the planet seemed to be a notch or depression. But this could be understood merely in terms of the darkness of the dark area: If it were dark against a dark sky, we would not see it at all. We might gain the mistaken impression of a notch or depression. The prevailing opinion of most astronomers seems to have been that the dark areas were slightly lower than the bright, but the difference was estimated by Lowell as only half a mile or less.

In 1966, I re-examined this problem with Dr. James Pollack. We used two main arguments. Mars has in its winter hemisphere a large polar cap which, at various times, has been ascribed to frozen water or frozen carbon dioxide. Even at the present time its composition is unsettled; both substances are probably present. As the polar cap retreats in each hemisphere, once each year, there are regions where frost is left behind. Later, when the frost leaves these regions, they are found to be brighter than their surroundings. By analogy with the Earth, we might expect them to be high mountainous regions that remain frosted after the snows of the valleys have melted or evaporated. Indeed, one Martian polar region—the so-called Mountains of Mitchel—was identified as mountainous by this argument alone.

But why are terrestrial mountains the last places to be frost-free? Because it is colder as we walk uphill, as every mountaineer knows. But why does it get colder as we walk uphill? Do the reasons that make terrestrial mountaintops colder than their bases apply to Mars?

We concluded that all factors that make it colder while walking uphill on the Earth are inoperative on Mars, mainly because of the very thin Martian atmosphere. But the winds on Mars should be higher at mountaintops than in valleys, as on Earth. This is not a conclusion from analogy, but is based on the appropriate physics. Therefore, we imagined that snows are removed by high winds preferentially from the mountains of Mars, and that the bright areas that retain frost on Mars are, therefore, low.

Our second line of attack was based on the radar observations of Mars, which began in the middle 1960s. There was one piece of evidence that immediately caught our attention. When the small central part of the radar beam was positioned directly on a dark area of Mars, only a very small fraction of the radar signal was returned to Earth. But when an adjacent bright area, on one side or the other of this dark region, was under the center of the radar beam, the reflection was much stronger. This could be understood if the dark area were either much higher or much lower than the adjacent bright area. From the preliminary radar evidence then available, we concluded that if it had to be one or the other of these two alternatives, the dark areas had to be systematically high on Mars. We concluded that major elevation differences existed on Mars, in some cases as much as ten miles between adjacent bright and dark areas. The large-scale slopes were at most only a few degrees—not a very steep grade—and both the elevation differences and slopes were comparable to those on Earth, although the elevations seemed to be greater than here. The notion that the deserts generally were lowlands seemed consistent with the notion of fine sand and dust being trapped in low valleys, with the tops of mountains—where the winds are higher—being scoured of small, bright, fine particles.

In the few years following our analysis many more detailed radar studies were done—principally by a group at the Haystack Observatory of the Massachusetts Institute of Technology, headed by Professor Gordon Pettengill. For the first time it was possible to do direct radar altitude measurements. Instead of using our indirect arguments, the technology had reached a point where it was possible to measure how long it took the radar signal to reach Mars and be returned from it. Those places on Mars from which the radar signal took longest to return were farthest from us, and, therefore, deepest. Those regions from which the radar signal took the least time to return were closer to us, and, therefore, highest. In this way the first topographic maps of selected regions of the Martian surface were constructed. The maximum elevation differences and slopes were just about what we had concluded by much more indirect means.

But dark areas did not appear to be systematically higher than bright areas. Pettengill and his colleagues found that a bright region of Mars called Tharsis appeared to be very high—perhaps the highest region sampled on the planet. A major Martian bright circular area called Hellas—Greek for "Greece"—indeed turned out to be very low from later nonradar observations. A somewhat similar feature called Elysium, also large and bright and roughly circular, turned out to be high. The darkest big Martian area, Syrtis Major, turned out to be a steep slope.

Why were Pollack and I only partly right? Because of Occam's Razor, a convenient and frequently used principle in science, but one that is not infallible. Occam's Razor recommends that, when faced with two equally good hypotheses, we choose the simpler. We had assumed that dark areas were either systematically high or systematically low. If that were the case, dark areas would have to be systematically high. But that is not the case; dark areas can be either high or low. Our conclusions only reflected our assumptions.

But I am very pleased that we were able, through logic and physics, to get the story at least partly right, and to demonstrate that there are enormous elevation differences on Mars, elevations much vaster than Lowell had expected. I find it more difficult, but also much more fun, to get the right answer by indirect reasoning and before all the evidence is in. It's what a theoretician does in science. But the conclusions drawn in this way are obviously more risky than those drawn by direct measurement, and most scientists withhold judgment until there is more direct evidence available. The principal function of such detective work—apart from entertaining the theoretician—is probably to so annoy and enrage the observationalists that they are forced, in a fury of disbelief, to perform the critical measurements.

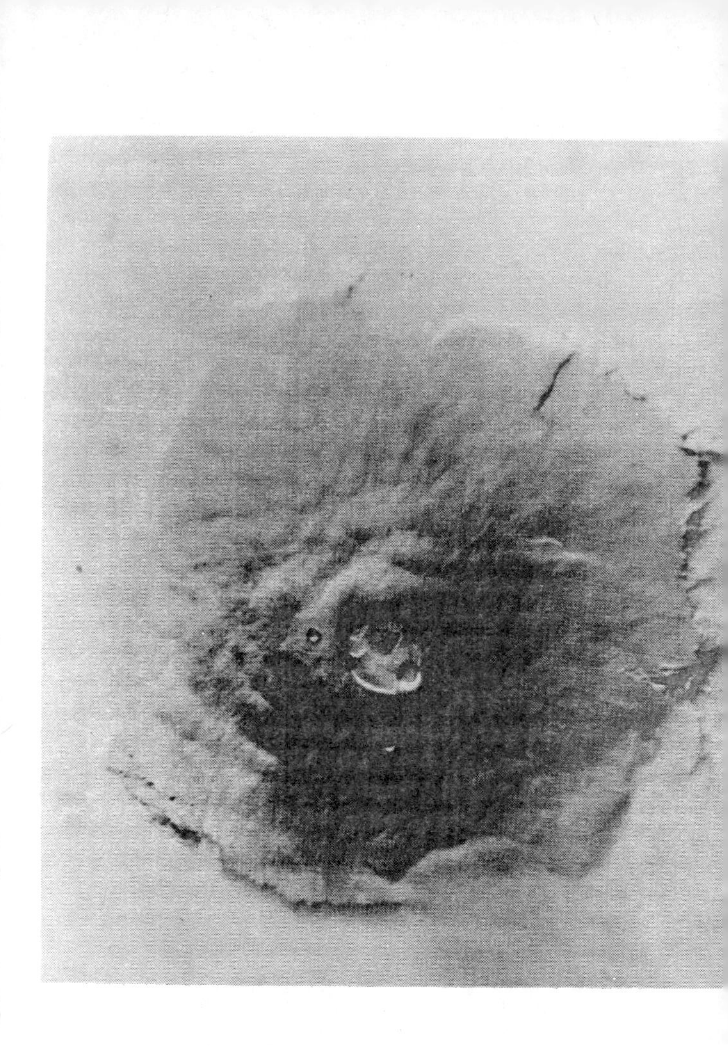

Mariner 9 mosaic of four photographs of the largest known volcano in solar system, Nix Olympica—seen vertically from above. Courtesy, NASA.

17.

The Mountains of Mars

II. OBSERVATIONS FROM SPACE

The epic flight of *Mariner* 9 to Mars in 1971 produced a new set of definitive and direct measurements concerning the mountains and elevations of Mars. Moderately complete elevation terrain maps of Mars have been developed as a result of the ultraviolet spectrometer, the infrared interferometric spectrometer, and the S-band occultation experiments aboard *Mariner* 9. But the most striking information on the mountains of Mars came from the television experiment.

The first pictures that *Mariner* 9 returned from Mars, obtained even before orbital insertion on November 14, 1971, showed an almost completely featureless planet. The south polar cap could be discerned dimly, but the bright and dark markings, which had been seen and debated for over a century, were nowhere to be found. This was not a failure of the television camera, but rather the result of a spectacular planetwide dust storm, which had begun in late September and would not significantly subside until early January.

The earliest pre-orbital pictures and the first few days of orbital pictures showed no significant nonpolar detail—except in the region of Tharsis. Here, there were four dark, somewhat irregular spots to be seen, three of them in an approximate straight line running northeast to southwest; the fourth was isolated away from them and to the west. Since there was otherwise nothing much visible on the planet, I devoted some attention to these spots in the early phases of the mission—so much attention that for a while they were known as "Carl's Marks" by several of my wittier co-investigators. I, in turn, proposed naming them Harpo, Groucho,

Chico, and Zeppo, but this was all before their significance was established.

The isolated spot corresponded in position quite well with the classical Martian feature named Nix Olympica—Greek for the Snows of Olympus, the home of the gods. The other three spots seemed to correspond to no familiar Martian surface features. But Bradford Smith, astronomer at New Mexico State University, pointed out that they corresponded quite well (as did Nix Olympica) to places on Mars that exhibited local afternoon brightening as observed from Earth. In some of Smith's ground-based telescopic photographs, obtained with a blue or violet filter and when there was no dust storm on Mars, these four places appeared as brilliant white spots, even though the contrast between the usual bright and dark areas was very small and the usual markings of Mars were indiscernible (the usual situation when Mars is viewed in blue or violet light rather than in orange or red light). Were we observing some sort of dark clouds in the midst of the dust storm at sites where bright clouds were usually found?

Another *Mariner 9* experimenter, William Hartmann of Science Applications, Inc., Tucson, Arizona, performed a computer contrast-enhancement of the original photographs of the four spots and found some faint indication of circular central regions in at least two of them. Indeed, the *Mariner 6* and 7 photographs of Nix Olympica, taken in 1969, showed a similar indication there.

By this time, the extent and severity of the dust storm had become evident, and part of our preplanned mission for *Mariner 9* to map the planet had to be postponed. This then freed a significant picture-taking ability for high-resolution, close-up photographs of the four spots. These experiments of opportunity were possible only because *Mariner 9* had a major adaptive capability. The scan platform, on which the cameras were located, could be aimed at many desired spots on Mars, and the technical staff of the Jet Propulsion Laboratory of the California Institute of Technology was able to change its plans quickly enough to accommodate the changed scientific needs of the mission. Because of the design of the spacecraft and the adaptability of its controllers, the first close-up photographs of the four spots began coming in.

Each spot had a vaguely circular center. There were parallel arcuate segments. There was a kind of scalloping. All these features were dark against a bright surrounding, corresponding to the dark appearance of the spots as seen initially in low resolution.

The particular shapes that we had seen in the early pictures held no particular significance for me. But I was struck by the fact that these circular features occurred in Tharsis, the highest region on Mars. These features were craters. Why were we seeing them and virtually no other Martian features? Because they must be the highest regions in Tharsis, a region already enormously elevated. The four spots, therefore, seemed to me to be vast mountains poking through the dust. I proposed that as time went on and the dust storm settled (from experience with other Martian global dust storms over decades of observation, we knew the dust storms would have to settle eventually), we would see more and more of these mountains, clear down to their bases. I even thought it possible that we could produce topographic maps from the sequence of emerging detail as the dust settled. Unfortunately, the settling out of the dust was a very irregular affair, and this suggestion has not yet borne fruit.

Geologist members of the *Mariner 9* television team, such as Harold Masursky and John McCauley, of the U. S. Geological Survey, were taken with the *form* of the craters, and quite early identified them—by analogy with similar features on Earth—as vast volcanic piles with summit calderas. I have always been mistrustful of arguments from terrestrial analogy. After all, Mars is quite another place. For all we knew—at least, for all *I* knew—quite different geological processes might operate there, and Earth-like features might be produced by different causes.

However, by another route I reached the same conclusion as the geologists: There are only two processes we know that produce craters—the impact of interplanetary debris (the origin, for example, of most of the craters on the Moon) and vulcanism. It would be asking too much to expect that the large meteorites or small asteroids that carved out four of the largest impact craters in Tharsis knew enough to land on the top of the four highest mountains in Tharsis. Much more plausible is the idea that the mechanism that made the mountain made the crater. That mechanism is called vulcanism.

As the dust storm cleared, the true magnitude of these four volcanic mountains became clear. The largest of them, Nix Olympica, is five hundred miles across, larger than the largest such feature on the Earth, the Hawaiian Islands. The altitudes of the spots have not yet been determined with precision, but they appear to be ten to twenty miles above the mean level of the planet. (We cannot talk of sea level on Mars because there are not—today, at any rate—any seas there.) Over a dozen smaller volcanoes have since been found in other regions of Mars.

The infrared radiometer on *Mariner 9* showed no sign of hot lava in the summit calderas of the craters. On the other hand, their fresh appearance and the almost total absence of meteorite cratering on their slopes show them to be very young objects, geologically speaking—probably no more than a few hundred million years old, possibly younger.

The association of clouds with these volcanic mountains could be due to contemporary outgassing from the calderas —steam, for example, being exhaled up volcanic vents. But it seems more likely that the clouds are present at the summits of these mountains precisely because these mountains are so high. An imaginary parcel of Martian air, rising along the slope of the mountain, expands and cools. (The air gets colder as we go upward in the Martian atmosphere. But because the air is so thin on Mars, it cannot exchange heat well with the surface; thus the *surface* does not get cold as we go uphill on Mars, as we discussed earlier.) When the temperature in the parcel of air drops below the freezing point of water, all of the water vapor in the parcel condenses out into ice crystals. The amount of water vapor we know to exist in the Martian atmosphere, the heights of the mountains, and the amount of small ice crystals necessary to produce a visible cloud together work out correctly for this to be the explanation of mountain clouds on Mars.

Recent vulcanism on Mars implies outgassing, whether or not the clouds that we see at the summits of these volcanoes are signs of outgassing. When hot lava flows to the surface, it carries with it a significant amount of gas—on the Earth, mainly water, but with a significant amount of other materials. Thus, the volcanoes that we see on Mars must have made an important contribution to the Martian atmosphere. In part, at least, the air has come out of these holes in the ground. Because Mars is so cold today, water can be trapped in many forms, such as ice, and not remain in the atmosphere. Much more gas could have been produced by these volcanoes than we see in the Martian atmosphere today. If there is life on Mars, it will almost surely be based on the exchange of material with the atmosphere—just as on Earth, where the cycle of green-plant photosynthesis and animal respiration is predominant. If there is life on Mars, these volcanoes may—at least indirectly—have played an important role in its present development.

After the dust storm cleared, *Mariner 9* was moved into a higher orbit, to facilitate the geological mapping originally planned. The spacecraft worked many times longer than its designers had expected. Complete geological coverage of the planet has been accomplished down to a resolution of half a mile. The resulting geological maps reveal an enormous array of linear ridges and grooves that surround the Tharsis Plateau—as if a third or a quarter of the whole surface of Mars were cracked in some colossal recent event that lifted Tharsis. The most spectacular of these quasilinear features is an enormous rift valley in a region called Coprates. It runs 80 degrees of Martian longitude and is almost exactly as long as the largest rift valley on the Earth, the East African Rift Valley, which runs up the entire east coast of Africa to the Dead Sea. Since Mars is a smaller planet, the Coprates Rift Valley is, relatively speaking, a much more impressive feature.

The East African Rift Valley occurs because of sea floor spreading and continental drift. The African and Asian continents are slowly moving away from each other, and the chasm that is developing there is the East African Rift Valley. But continental drift is thought to be due to the slow circulation of material in the mantle of the Earth. Should we then conclude that Mars, despite its smaller size and lower internal temperatures, also has mantle convection and continental drift? Or is it possible that different processes produce similar features on the two planets?

Whatever the answer, we cannot help but learn a great deal more about the old Earth-bound science of geology—with its practical future disciplines, such as earthquake prediction and control—by examining the geology of our neighboring planet Mars.

18.
The Canals of Mars

In 1877 (as in 1971) the planet Mars was close—forty million miles from Earth. European astronomers, with newly developed telescopes, prepared for what was then Man's most detailed look at our planetary neighbor. One of them was Giovanni Schiaparelli, an Italian observing in Milan and a collateral relative of the present couturier and perfume enterpriser.

Generally speaking, the telescopic view of Mars was blurred and fuzzy, interrupted by the variable turbulence in the Earth's atmosphere that astronomers call "seeing." But there were moments when the Earth's atmosphere steadied and the true detail on the disc of Mars seemed to flash out. Schiaparelli was astonished to see a network of fine straight lines covering the disc of Mars. He called these lines *canali*, which in Italian means "channels." However, *canali* was translated into English as "canals," a word with a clear imputation of design.

Schiaparelli's observations were taken up by Percival Lowell, a diplomat once posted in Chosen, the present Korea. A Boston Brahmin, the brother of the president of Harvard University and of an even more famous personage, the poetess Amy Lowell (for some reason renowned for smoking little black cigars), Lowell established a private observatory in Flagstaff, Arizona, to study the planet Mars. He found the same *canali* that Schiaparelli had. He extended their description and elaborated an explanation.

Mars was, Lowell concluded, a dying world on which intelligent life had arisen and accommodated itself to the perils of the planet. The chief peril was the dearth of water. The

Martian civilization, Lowell imagined, had constructed an extensive network of canals to carry water from the melting polar caps to the habitations in more equatorial climes. The turning point of the argument was the straightness of the canals, some of them following great circles for thousands of miles. Such geometrical configurations, Lowell thought, could not be produced by geological processes. The lines were too straight. They could only have been produced by intelligence.

This is a conclusion with which we all can agree. The only debate is about which side of the telescope the intelligence was on. Lowell believed that the penchant for Euclidian geometry was on the distant end of the telescope. But the difficulties in drawing a great deal of mottled fine detail in a few seconds of good seeing are so great that the eye-brain-hand combination is sorely tempted to connect such disconnected features into straight lines. Many of the best visual astronomers observing Mars between the turn of the century and the dawn of the space age found that, while they could see canals under conditions of good but not superb seeing, they were able in the extremely rare moments of perfect seeing to resolve the straight lines into a multitude of spots and irregular detail.

Then it was found that at least the vast bulk of the polar caps are carbon dioxide and not frozen water. The atmospheric pressure was discovered to be much less than on Earth. Liquid water was found to be entirely impossible. The idea of advanced forms of life and canals on Mars died. And yet . . .

As the planet-wide dust storm cleared in 1971, the *Mariner* 9 spacecraft began to photograph a region called Coprates by the classical observers. Coprates was one of the largest *canali* found by Lowell, Schiaparelli, and their followers. Toward the end of the dust storm, Coprates was revealed to be an enormous rift valley running three thousand miles east to west near the Martian equator, fifty miles wide in spots and a mile deep. It was not perfectly straight—it was certainly not an engineering work; but it was a vast gash proportionately longer than any such feature on Earth.

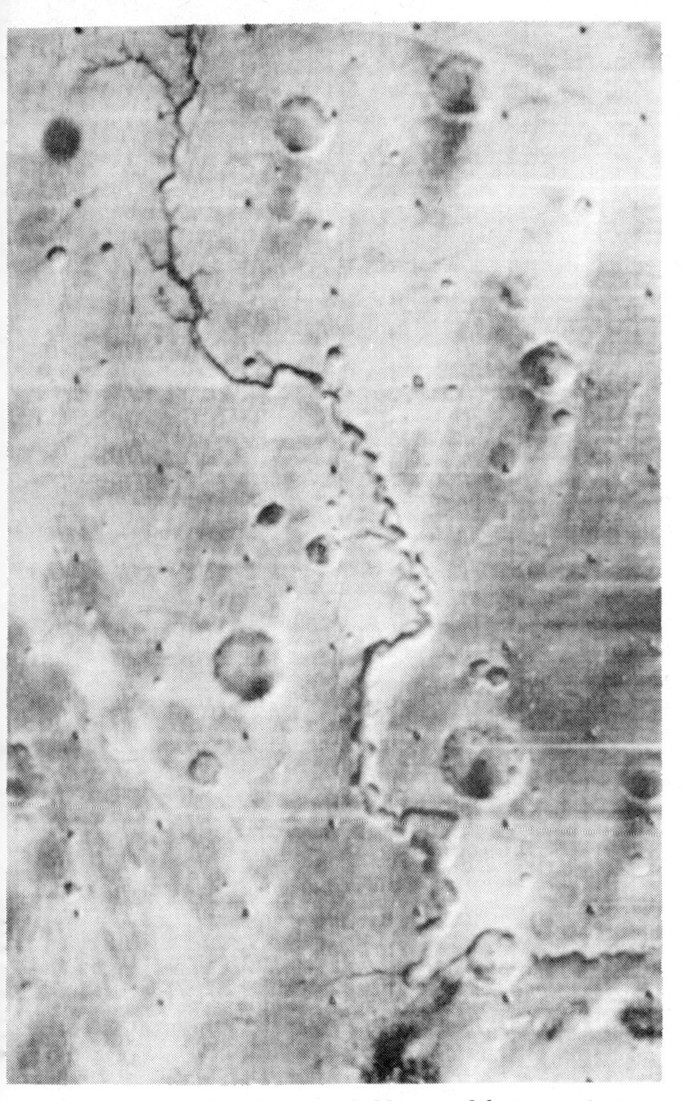

A sinuous channel on Mars—probably carved by an ancient river. Courtesy, NASA.

And running out of Coprates were features that were very curious indeed—sinuous channels, meandering through the highlands above the Coprates Valley and graced with beautiful little tributaries. If such channels had been seen on Earth, they would unhesitatingly have been attributed to running water. But on Mars the surface pressures are so low that liquid water would instantly vaporize, just as the pressures on Earth are so low that liquid carbon dioxide vaporizes instantly. On Earth we have solid carbon dioxide and gaseous carbon dioxide, but not liquid carbon dioxide. On Mars this absence of the liquid phase is true as well for water.

But as the *Mariner 9* photographic mission continued, a variety of additional channels were discovered: Channels with second- and third-order tributary systems, channels without a crater at their beginning or end, channels with teardrop-shaped islands in their midst, channels with braided termini, like those cut on Earth by episodic flooding.

There seems to be little doubt that most of the several dozen longest such channels (the longest are hundreds of miles long), and hundreds of smaller ones, were cut by running water. But since there can be no liquid water on Mars today, the channels must have been cut in a previous epoch of Martian history—when the total pressures were larger, the temperatures higher, and the availability of water greater.

The channels revealed by *Mariner 9* speak eloquently of the possibility of massive climatic change on Mars. In this view, Mars is today in the throes of an ice age, but in the past—no one knows just how long ago—it possessed much more clement and Earth-like conditions.

The reasons for such dramatic climatic changes are still being hotly debated. Before the *Mariner 9* launch, I proposed that such climatic changes leading to episodes of liquid water might occur on Mars. They might be driven by the precession of the equinoxes, a well-known motion akin to the slow, drifting precession of a rapidly spinning top. The precessional periods on Mars are something like fifty thousand years. If we are now in a precessional winter, with an extensive North polar ice cap, twenty-five thousand years ago may have been

The great Coprates Rift Valley. Courtesy, NASA.

the precessional winter with an extensive polar ice cap in the South.

But twelve thousand years ago may have been the epoch of precessional spring and summer. The dense atmosphere of that time is now locked away in the polar caps. Twelve thousand years ago may have been a time on Mars of balmy temperatures, soft nights, and the trickle of liquid water down innumerable streams and rivulets, rushing out to join mighty, gushing rivers. Some of these rivers would have flowed into the great Coprates Rift Valley.

If so, twelve thousand years ago was a good time on Mars for life similar to the terrestrial sort. If I were an organism on Mars, I might gear my activities to the precessional summers and close up shop in the precessional winters—as many organisms do on Earth for our much shorter annual winters. I would make spores; I would make vegetative forms; I would go into cryptobiotic repose; I would hibernate until the long winter had subsided. If this is indeed what Martian organisms do, we may be arriving at Mars twelve thousand years too early—or too late!

But there is a way to test these ideas. One way the hypothetical Martian organisms would know that the precessional spring has arrived is by the reappearance of liquid water. Therefore, as Linda Sagan has mentioned, the recipe for detecting life on Mars is "Add water." And this is just what the U. S. Viking biology experiments, scheduled to land on Mars in 1976 and search for microbes, will do. An automatic arm will drop two samples of Martian soil into liquid water. A third sample will be inserted into a chamber with no liquid water. If the first two experiments give positive biological results, and the third experiment does not, some support will be given to this idea that Martian organisms are waiting out the long winter.

But it is entirely possible that the designs of these experiments have been too Earth-chauvinist. There may be Martian organisms that enjoy the present environment and are drowned in liquid water. The idea of Martian organisms as sleeping beauties, awaiting a somewhat wet kiss from Viking, is a long shot—but a fascinating one.

By no means do all of the channels correspond to the positions of the classical *canali* drawn by Lowell and Schiaparelli. Some, like Ceraunius, appear to be ridges. Others correspond to no detail that can now be made out. But some, like Coprates, are grooves in the Martian terrain. There *are* channels on Mars. They may have biological implications, of a different sort than Lowell imagined (as the long-winter model suggests), or they may have no connection with Martian biology at all.

The canals of Lowell do not exist, but the *canali* of Schiaparelli are there to be seen, more or less. One day in the future, perhaps, the channels will again be filled with water and, for all we know, with visiting gondoliers from the planet Earth.

19.

The Lost Pictures of Mars

The *Mariner* 9 mission to Mars radioed back to Earth 7,232 photographs that revolutionized our knowledge about the planet. Many hundreds of these pictures were devoted to studying variable features, the time changes in the relative configurations of bright and dark markings on the surface of the planet now known to be due largely to wind-blown dust. We have found thousands of bright and dark streaks, beginning in local impact craters and stretching across tens of miles of Martian surface. They point in the direction of the prevailing winds. We think they are produced by high winds carrying dust out of the craters and depositing it on the surface beyond the crater ramparts. These streaks are natural wind-direction indicators and, perhaps, anemometers laid down on the Martian surface for our edification and delight. We have discovered dark irregular patches or splotches, mostly residing in the interiors of craters, which tend to lie on the leeward walls of the craters. Thus, the splotches as well as the streaks are wind indicators. Some of the splotches have been resolved by *Mariner* 9 into enormous fields of parallel sand dunes.

We have discovered many cases of dark streaks and splotches varying through the mission in outline or extent. The positions and variabilities of these dark features correspond well to the classical dark markings of Mars, observed by ground-based astronomers for more than a century and most often attributed by them to seasonally changing dark vegetation on the Martian surface. But our *Mariner* 9 evidence points unambiguously to a meteorological, rather

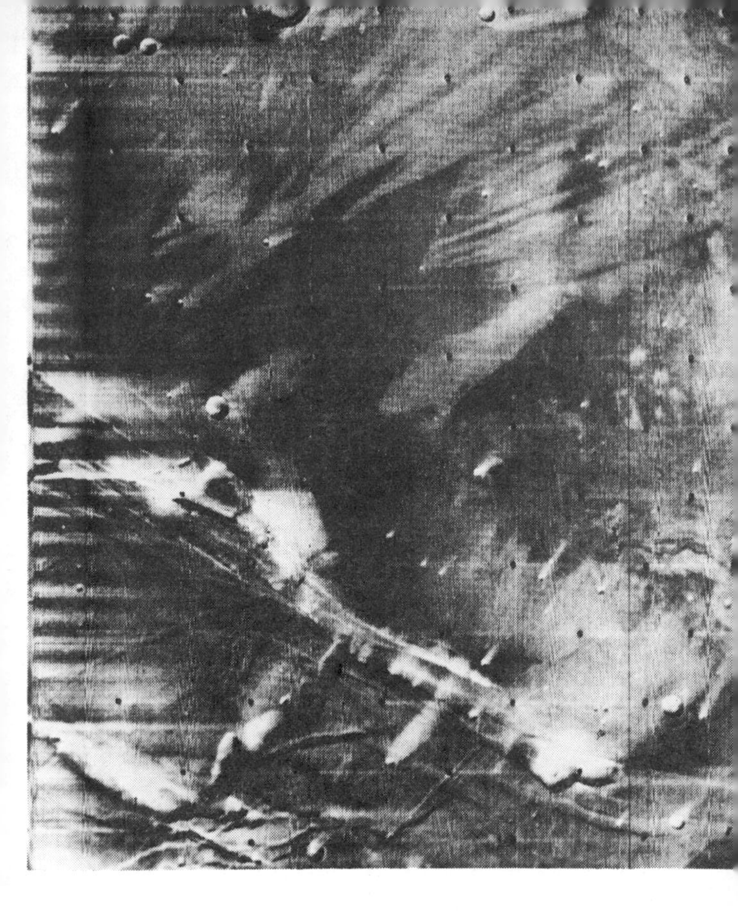

Three *Mariner 9* photographs showing effects of high winds near the Martian surface. Above: Wind streaks emanating from impact craters. Each streak is a thin layer of wind-blown dust. Top right: Splotches in crater interiors. Some of these features are dark rock exposed when overlying dust is removed by high winds. Bottom right: Resolution of one of the splotches in top right into an array of longitudinal sand dunes, comparable in dimensions to the largest such dune fields on Earth. Courtesy, NASA.

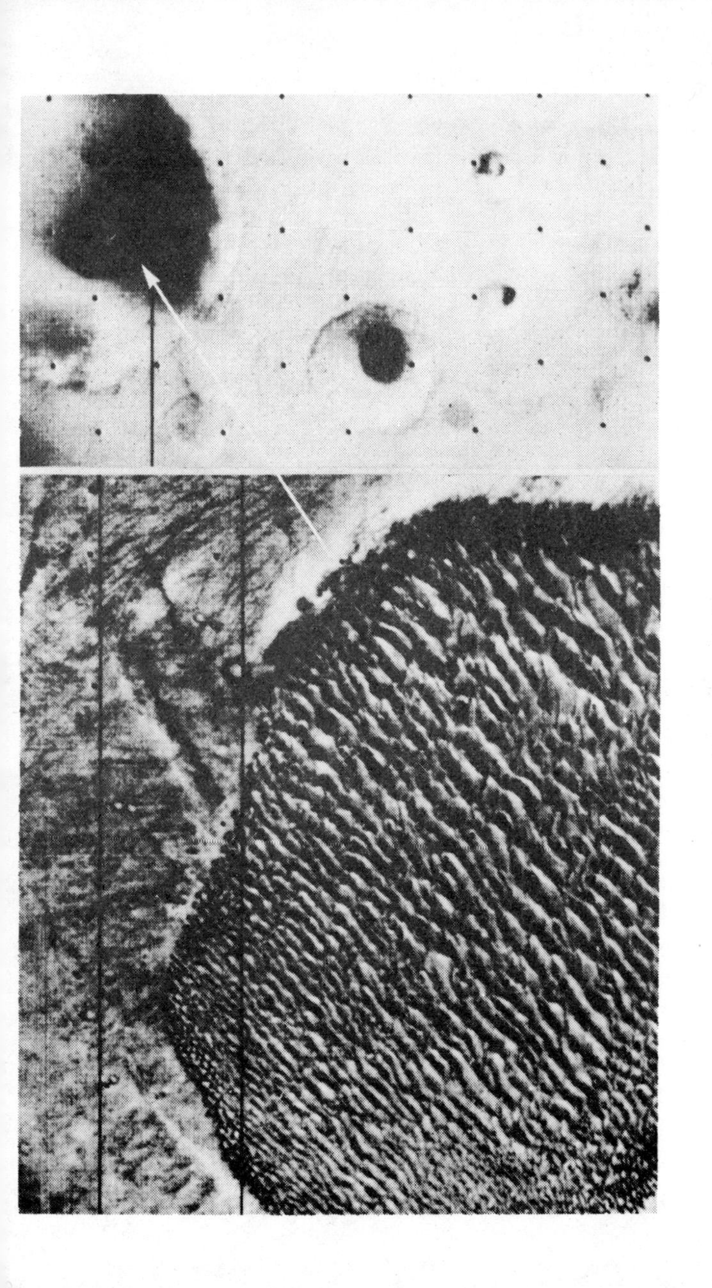

than a biological, explanation of the Martian seasonal changes.

This in no way excludes life on Mars. It merely means that if there is life on Mars, it is not easily detectable over interplanetary distances. The same is true in reverse: *Photographic* detection of life on Earth in daylight from the vantage point of Mars is impossible, as we have found by studying several thousand orbital photographs of our own planet. But the time-varying streaks and splotches on the Martian surface are a new and most exciting Martian phenomenon, which cries out for further study.

Since the Martian changes occur slowly, the variable-features objectives required very long time intervals between two pictures of the same region to see what changes had occurred. At the very end of the mission, fifteen photographs were successfully taken by the *Mariner 9* cameras of regions in Syrtis Major and Tharsis, important for understanding the long-term variations. But when the time came to point the high-gain antenna of *Mariner 9* to the Earth, so that these pictures could be transmitted by playing back the spacecraft's tape recorder, the last of the attitude-control gas was used up, Earth-lock could not be acquired, and playback did not occur. The spacecraft had literally run out of gas.

About a year before the *Mariner 9* mission was launched, the possibility was raised that the spacecraft would run out of control gas. A solution was proposed: That the propulsion tanks be connected to the attitude-control gas system—a kind of spacecraft anastamosis. Excess propulsion gas could then be used for attitude control in case the attitude-control nitrogen was exhausted. This possibility was rejected—largely because of its expense. It would have cost $30,000. But no one expected *Mariner 9* to last long enough to use up its attitude-control gas. Its nominal lifetime was ninety days—and it lasted almost a full year. The engineers had been overly conservative in assessing their superb product.

In retrospect, it sounds very much like false economy. With an adequate supply of attitude-control gas, the spacecraft might have lasted another full year in orbit around Mars. About $150 million of science might have been bought for

$30,000 of pipe. Had we known that the spacecraft would die from a lack of nitrogen, I am almost certain that the planetary scientists involved would have raised the $30,000 themselves.

In fact, there are many such critical junctures in the space program where the addition of only a small amount of money can greatly increase the scientific return from a given mission. But NASA, severely limited by funding limitations imposed by Congress, the White House, and the Office of Management and Budget, has not had such small increments of money. If it were possible, and if a generous donor could be found, this would be a superb use for private philanthropy.

But these are idle musings. No anastamosis was performed; the final playback was not accomplished. Sitting there still on the *Mariner* 9 tape recorder are fifteen vital photographs of the planet. They will never be returned under *Mariner* 9's own power. It has now also lost solar lock; sunlight is no longer being converted to electricity on its four great solar panels, and there is no way to reactivate it. We may never know what Tharsis and Syrtis Major looked like around the beginning of November 1972 from the vantage point of Martian orbit.

Or perhaps we will. *Mariner* 9 is in an orbit that is slowly decaying in the Martian atmosphere. But the decay is so slow that the spacecraft will not crash into Mars for another half century. Long before then there should be manned orbital flights around Mars. Rendezvous and docking maneuvers are reasonably well developed in manned missions even now. Perhaps, then, sometime around 1990, as a small side-trip in a grand manned-orbital exploration of Mars, there will be a rendezvous with *Mariner* 9. The old and battered spacecraft will be taken aboard a large manned station and returned home—perhaps to be put in the Smithsonian Institution; perhaps to prevent terrestrial micro-organisms on *Mariner* 9 from reaching Mars; but perhaps, also, to rescue and read off the fifteen lost pictures of the *Mariner* 9 mission.

The extremes of climate on the Earth. Top: Dune field in the Sahara from *Sahara*, C. Krüger, ed., New York, Putnam, 1969. Bottom: Crevices in the Vatnajökull Glacier in Iceland, from *The World from Above* by O. Bihalji-Merin, Hill and Wang, New York, 1966.

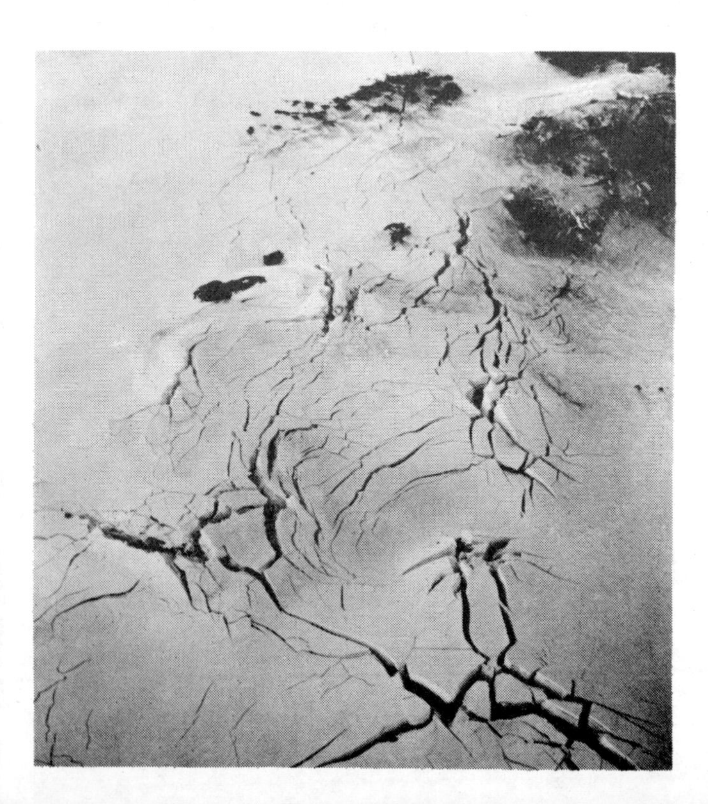

20.

The Ice Age and the Cauldron

On our tiny planet, spinning in an almost circular orbit at a nearly constant distance from our star, the climate varies, sometimes radically, from place to place. The Sahara is different from the Antarctic. The Sun's rays fall directly on the Sahara and obliquely on the Antarctic, producing a sizable temperature difference. Hot air rises near the equator, cold air sinks near the poles—producing atmospheric circulation. The motion of the resulting air current is deflected by Earth's rotation.

There is water in the atmosphere, but when it condenses, forming rain or snow, heat is released into the atmosphere, which in turn changes the motion of the air.

Ground covered by freshly fallen snow reflects more sunlight back to space than when it is snow-free. The ground becomes colder yet.

When more water vapor or carbon dioxide is put into the atmosphere, infrared emission from the surface of the Earth is increasingly blocked. Heat radiation cannot escape from this atmospheric greenhouse, and the Earth's temperature rises.

There is topography on Earth. When wind currents flow over mountains or down into valleys, the circulation changes.

At one point in time on one tiny planet, the weather, as we all know, is complex. The climate, at least to some degree, is unpredictable. In the past there were more violent climatic fluctuations. Whole species, genera, classes, and families of plants and animals were extinguished, probably because of climatic fluctuations. One of the most likely explanations of the extinction of the dinosaurs is that they were large animals with poor thermoregulatory systems; they were unable to burrow, and, therefore, unable to accommodate to a global decline in temperature.

The early evolution of man is closely connected with the emergence of the Earth from the vast Pleistocene glaciation. There is an as yet unexplained connection between reversals of the Earth's magnetic field and the extinction of large numbers of small aquatic animals.

The reason for these climatic changes is still under serious debate. It may be that the amount of light and heat put out by the Sun is variable on time scales of tens of thousands or more years. It may be that climatic change is caused by the slowly changing direction between the tilt of the Earth's rotational axis and its orbit. There may be instabilities connected with the amount of pack ice in the Arctic and Antarctic. It may be that volcanoes, pumping large amounts of dust into the atmosphere, darken the sky and cool the Earth. It may be that chemical reactions reduced the amount of carbon dioxide and other greenhouse molecules in the atmosphere, and the Earth cooled.

There are, in fact, some fifty or sixty different and, for the most part, mutually exclusive theories of the ice ages and other major climatic changes on Earth. It is a problem of substantial intellectual interest. But it is more than that. An understanding of climatic change may have profound practical consequences—because Man is influencing the environment of the Earth, often in ways poorly thought-out, ill-understood, and for short-term economic profit and individual convenience, rather than for the long-term benefit of the inhabitants of the planet.

Industrial pollution is churning enormous quantities of foreign particulate matter into the atmosphere, where they are carried around the globe. The smallest particles, injected into the stratosphere, take years to fall out. These particles increase the albedo or reflectivity of Earth and diminish the amount of sunlight that falls on the surface. On the other hand, the burning of fossil fuels, such as coal and oil and gasoline, increases the amount of carbon dioxide in the Earth's atmosphere which, because of its significant infrared absorption, can increase the temperature of the Earth.

There is a range of effects pushing and pulling the climate in opposite directions. No one fully understands these interactions. While it seems unlikely that the amount of pollution currently deemed acceptable can produce a major climatic change on Earth, we cannot be absolutely sure. It is a topic worth serious and concerted international investigation.

Space exploration plays an interesting role in testing out theories of climatic change. On Mars, for example, there are periodic massive injections of fine dust particles into the atmosphere; they take weeks and sometimes months to fall out. We know from the *Mariner 9* experience that the temperature structure and climate of Mars are severely changed during such dust storms. By studying Mars, we may better understand the effects of industrial pollution on Earth.

Likewise for Venus. Here is a planet that appears to have undergone a runaway greenhouse effect. A massive quantity of carbon dioxide and water vapor has been put into its atmosphere, so cloaking the surface as to permit little infrared thermal emission to escape into space. The greenhouse effect has heated the surface to 900 degrees F or more. How did this greenhouse-overkill happen on Venus? How do we avoid its happening here?

Study of our neighboring planets not only helps us to generalize the study of our own, but it has the most practical hints and cautionary tales for us to read—if only we are wise enough to understand them.

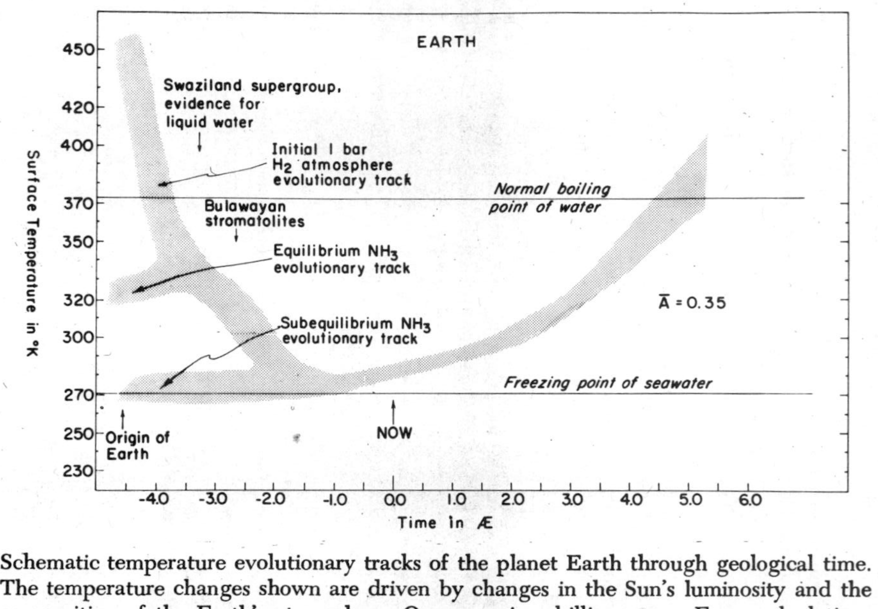

Schematic temperature evolutionary tracks of the planet Earth through geological time. The temperature changes shown are driven by changes in the Sun's luminosity and the composition of the Earth's atmosphere. One aeon is a billion years. From calculations by the author. Courtesy, American Association for the Advancement of Science.

21.

Beginnings and Ends of the Earth

Stars, like people, do not live forever. But the lifetime of a person is measured in decades, the lifetime of a star in billions of years.

A star is born out of interstellar clouds of gas and dust. For a while, it stably converts hydrogen to helium in the thermonuclear furnaces of its deep interior. Then, in stellar old age, it encounters a set of minor or major catastrophes—a slow trickle or an explosive injection of star-stuff into space. During the more or less stable portion of the lifetime of the star, the hot interior region, converting hydrogen into helium, gradually eats its way outward from the very center. In the course of time, the star becomes slowly, almost imperceptibly, brighter.

After the flares and other impetuosities of its early adolescence, our Sun settled down to a more or less constant radiation output. But four billion years ago it was about 30 percent dimmer than it is today. If we assume that four billion years ago the Earth had the same distribution of land and water, clouds and polar ice, so that it absorbed the same relative amount of sunlight as it does today, and if we also assume that it had the same atmosphere as it does today, we can calculate what its temperature would have been. The calculation reveals a temperature for the entire Earth significantly below the freezing point of seawater. In fact, even two billion years ago, under these assumptions, the Sun would not have been bright enough to keep the Earth above the freezing point.

But we have a wide variety of evidence that this was not the case. There are in old mud deposits ripple marks caused by liquid water. There are pillow lavas produced by undersea volcanoes. There are enormous sedimentary deposits that can only be produced on ocean margins. There are biological products, called algal stromatolites, which can only be produced in water.

So what is wrong? Either our theory of the evolution of the Sun is wrong or our assumption that the early Earth is like the present Earth is wrong. The theory of solar evolution seems to be in good shape. What uncertainties exist do not appear to affect the question of the Sun's early luminosity.

The most likely resolution of this apparent paradox is that something was different on the early Earth. After studying a wide range of possibilities, I conclude that what was different, two billion years ago and earlier, was the presence of small quantities of ammonia in the Earth's atmosphere. Ammonia is present on Jupiter today; it is the form of nitrogen expected under primitive conditions. It absorbs very strongly at the infrared wavelengths that the Earth likes to emit to space. Ammonia on the primitive Earth would have held heat in, increasing the surface temperature through the greenhouse effect and keeping the global temperature of Earth at congenial levels—for the origin and early history of life and for liquid water to have been abundant early in the history of the planet. And ammonia is one of the atmospheric constituents needed for making the building blocks of life. The study of the Sun's evolution leads us to information about the early history, chemical composition, and temperature of the Earth, and, therefore, to the circumstances of its habitability. Stellar and biological evolution are connected.

What about the future evolution of the Sun? The Sun is steadily growing brighter. About four billion years from now the Sun will be sufficiently brighter that there will be a greatly enhanced runaway greenhouse effect on Earth, just as there is today on Venus. Our oceans will boil, and carbon dioxide, now present as carbonates in the sedimentary rocks, will pour out into the atmosphere. The Earth will be an uninhabitable cauldron.

It is conceivable that the technology of those remote times will be equal to the task of preventing such a runaway, but it will be an extremely difficult engineering job. However, remarkably, the same increase in the brightness of the Sun some billions of years from now will convert Mars from a place where the average temperature is 100 degrees F below zero to a place that has temperatures almost exactly the same as those on Earth today.

When the Earth becomes uninhabitable, Mars will gain a balmy and clement climate. Our remote descendants, if any, may wish to take advantage of this coincidence.

22.

Terraforming the Planets

Both subtly and profoundly, the activities of life have affected the environment of our planet. Our atmosphere is composed of 20 percent oxygen and 80 percent nitrogen. The oxygen is produced almost entirely by green-plant photosynthesis. Similarly, the most recent evidence suggests that nitrogen is almost entirely a product of the biological activity of soil micro-organisms, which convert nitrates and ammonia into the gas N_2, molecular nitrogen. Not only are the principal constituents of our atmosphere closely controlled by biological activities, but the minor constituents are as well. To a significant extent, carbon dioxide is also buffered by the photosynthesis/respiration feedback loop. Even so minor a constituent of the Earth's atmosphere as methane, CH_4, is of biological origin.

In fact, life on Earth, invisible to photography, could be detectable with a small telescope and an infrared spectrometer from the vantage point of Mars. The Martians, if any, could easily observe, at a wavelength of 3.33 microns in the infrared, a strong absorption feature that straightforward analysis would reveal to be due to one part per million of methane in the terrestrial atmosphere. It should not be difficult to deduce that the methane is probably of biological origin. Methane is chemically unstable in an excess of oxygen. It is oxidized rapidly to carbon dioxide:

$$CH_4+2O_2=CO_2+2H_2O.$$

The amount of methane that would be in equilibrium with the great excess of oxygen in our atmosphere is less than one billion billion billionth the amount actually observed. How can this be? Methane must be produced at a rate so rapid that there is not time enough for oxygen to reduce its abun-

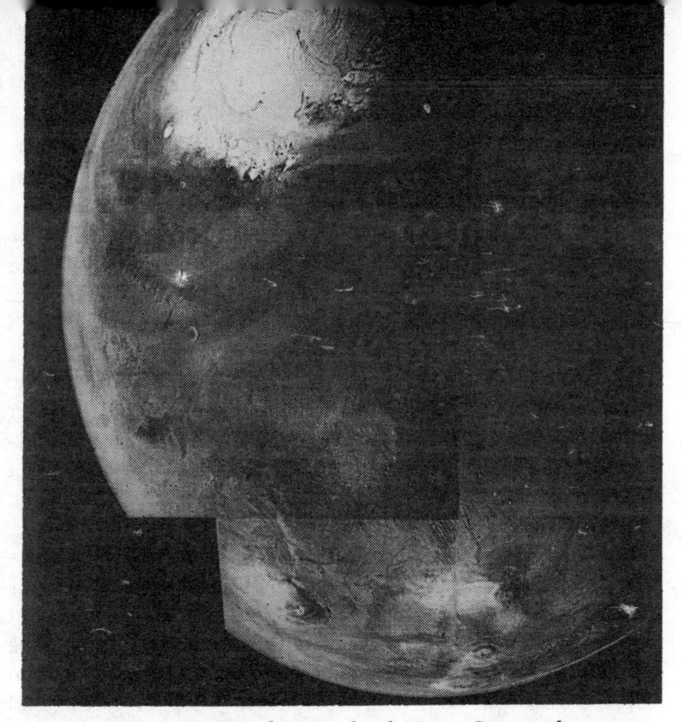

Mariner 9 composite photograph of Mars. Great volcanoes are seen at the bottom, the northern polar cap at the top. The amount of carbon dioxide and water frost locked in the polar cap, if released into the atmosphere, would probably produce much more Earth-like climatic conditions. Courtesy, NASA.

dance to the equilibrium amount. It might be that there are massive outpourings of methane from ancient petroleum fields on Earth. But because of the huge output required, this is a very unlikely hypothesis. It is far more likely that methane is produced by a biological process.

And this is indeed the case. There seems to be a debate in the ecological literature on two possible sources of this methane. One source is methane bacteria, which live in swamps and marshes—hence the term "marsh gas" to refer to methane. The principal other habitat of methane bacteria is in the rumens of ungulates. There is at least one school of ecological thought that believes that more methane is produced from the latter source than from the former. This means that bovine flatulence—the intimate intestinal activities of cows, reindeer, elephants, and elk—is detectable over interplanetary distances, while the bulk of the activities of mankind are invisible. We would not ordinarily consider the flatulence of cattle as a dominant manifestation of life on Earth, but there it is.

Inadvertently, with no conscious effort by mankind, life on Earth has reworked the environment in a major way. Through the effect of atmospheric pressure and composition on the climate, there is a feedback loop in which the climate itself may to some degree be controlled by the gas exchange reactions in which the life forms on Earth engage. In a way, life on Earth has terraformed Terra. It has to some extent made the Earth the way it is.

Is it possible that at some time in the future we might be able similarly to terraform other planets, to convert a Mars or Venus, today inhospitable to Man, into a clement and habitable environment? Such a change, if possible at all, should be done only after the most careful and responsible examination of the consequences. We would first want to understand thoroughly the present environment of the planet before altering it. We must scrupulously guarantee that any indigenous organisms on the planet would not be disrupted by terraforming. If Mars, for example, has a population of indigenous organisms that would be extinguished by terraforming, we should never perform such terraforming. But if the planet is lifeless, or if the organisms survive better under conditions closer to our own, it might be reasonable at some time in the future to consider such an alteration of a planetary environment.

Our motivations for planetary re-engineering must be clear. This is not a solution to the overpopulation problem. Several hundred thousand people are born every day on Earth. There is certainly no prospect in the immediate future of transshipping hundreds of thousands of people to other planets each day. In its entire history mankind has managed to launch one dozen people to another celestial body. Nor are we likely to see in the immediate future a thriving mining industry in which ores are extracted from another planet and transshipped to Earth: The freightage would be prohibitive.

And yet the human spirit is expansive; the urge to colonize new environments lies deep within many of us. Such activities can be performed without cosmic imperialism, without the kind of arrogance that characterized the European colonization of the New World, or the encroachment on the Indians in the settling by whites of the American West. Interplanetary colonization can be consistent with the highest aspirations and goals of mankind.

How would we do it? In the case of Venus, as we saw in Chapter 12, there is a crushing atmosphere, composed largely of carbon dioxide, and a searing surface temperature in excess of 900 degrees F. It would seem to be a formidable task indeed to convert this environment into one in which men could live and work without enormous technological assistance. But there is a bare possibility of re-engineering Venus into a quite Earth-like place, a possibility I suggested with some caution in 1961. The method assumes that the high surface temperature is produced by a greenhouse effect involving carbon dioxide and water, a conjecture that is much more plausible now than it was then. The idea is simply to seed the clouds of Venus with a hardy variety of algae—a genus called Nostocacae was suggested—which would perform photosynthesis in the vicinity of the clouds. Carbon dioxide and water would be converted into organic compounds, largely carbohydrates, and oxygen. The algae would, however, be carried by the atmospheric circulation down to deeper and hotter levels in the Venus atmosphere, where they would be fried. Frying an alga releases simple carbon compounds, carbon, and water into the atmosphere. The water content of the atmosphere thus remains fixed, and the net result is the conversion of carbon dioxide into carbon and oxygen.

The present greenhouse effect on Venus is due largely to carbon dioxide and water. The present total pressure on Venus is about ninety times that on the surface of Earth. The Venus atmosphere is largely composed of carbon dioxide. As the carbon dioxide is converted into carbon and oxygen, and the oxygen is chemically combined with the crust of

Venus, the total pressure would decline, decreasing atmospheric infrared absorption, reducing the greenhouse effect, and lowering the temperature.

It is possible, therefore, that the injection of appropriately grown algae into the clouds of Venus, algae able to reproduce there faster than they are fried, would in time convert the present extremely hostile environment of Venus into one much more pleasant for human beings.

The amount of water vapor in the Venus atmosphere, if condensed on the surface of the planet, would give a layer of water about one foot high—not an ocean, but enough to do irrigation and to provide for other human needs. It is also possible that water is available bound to the rocks on the surface of the planet.

No one can estimate whether this is a very likely scenario, or how long it would take to re-engineer the second planet from the Sun. It is perfectly possible that there is some flaw in the idea. For example, the high surface temperature may not be due to a greenhouse effect, but I think this is unlikely.

In any case, I think terraforming Venus is not impossible. The Nostoc scheme is an example of how human technology and science may, in periods quite short compared to geological time, rework the environment of another planet.

For Mars, as we saw in Chapter 18, there is now evidence that in comparatively recent times conditions on that planet were much more Earth-like than they are today. We mentioned the likelihood that enormous quantities of carbon dioxide and water are locked in the Martian polar caps, trapped as permafrost and chemically bound to the surface material elsewhere on the planet. Much of this CO_2 and H_2O may be released from the polar caps into the atmosphere twice each precessional cycle of fifty thousand years. Drs. Joseph Burns and Martin Harwit of Cornell University have considered a variety of technological schemes to induce more clement conditions on Mars hundreds of years from now, rather than thousands. These schemes involve alteration of the orbits of the Martian satellites or of a nearby asteroid to change the precessional motion of the planet, or the installation of an enormous orbiting mirror over the polar cap to melt the material frozen there. Even easier, however, might be to sprinkle carbon black over the caps, heat up the poles, increase the atmospheric pressure, and warm the planet.

Again, we do not know that such schemes will work, but they do not seem extremely impractical. It may very well be that on time scales of hundreds of years we will have the capability of converting Mars into a much more Earth-like planet than it would otherwise be.

The Moon and the asteroids are much less hospitable than Mars and Venus. They are so much less able to retain an atmosphere that the terraforming schemes we have been discussing are inapplicable to them. But even on airless worlds, the establishment of human colonies on their surfaces or even—in the case of small asteroids—in their interiors seems a possible future project for mankind. Such colonies would be much more constrained than those on a re-engineered Mars or Venus, and would require much greater attention to the husbanding of scarce resources.

Such colonies would be tenable only if significant natural resources—particularly frozen or chemically bound water—were to be found. In the case of the very surface of the Moon, the samples returned by Apollo astronauts showed virtually no such water at all. But it is entirely possible that large stores of water exist in cold, shadowed regions near the lunar poles or at substantial depths beneath the lunar surface.

It is not unlikely that on a time scale of a few centuries there will be extensive human colonies throughout the inner part of the Solar System and on some of the major satellites of the Jovian planets. The prospect is, of course, a difficult one; the engineering tasks are immense and the need to retain ecological respect for other environments pervasive. The danger of both forward and backward biological contamination must always be examined scrupulously.

There may even come a day when we shall be called to account for our stewardship of the Solar System. From that vantage point our own epoch will be viewed as a moment when we first left the cradle of our species and began, in a groping and tentative way, to explore and transform the space surrounding us.

A space caravel. Picture by Jon Lomberg after a drawing by Brueghel.

23. The Exploration and Utilization of the Solar System

At the very beginning of the twentieth century competent scientific and lay opinion held that airplanes were impossible. The end of the century, barring the dark specter of nuclear or ecological catastrophes, will probably see joint Soviet and American manned space expeditions to the nearer planets.

This is the century in which some of the oldest dreams of Man have been realized, in which mankind has sprouted wings and realized the aspirations of Daedalus and da Vinci. Air-breathing, man-carrying machines now circumnavigate our planet in less than a day; other machines, skimming above the atmosphere, carry men around our globe in ninety minutes.

There is a generation of men and women for whom, in their youth, the planets were unimaginably distant points of light, and the Moon was the paradigm of the unattainable. Those same men and women, in middle life, have seen their fellows walk upon the surface of the Moon; in their old age, they will likely see men wandering along the dusty surface of Mars, their journeys illuminated by the battered face of Phobos. There is only one generation of humans in the ten-million-year history of mankind that will live through such a transition. That generation is alive today.

This is also the moment in our history when, for the first time, the whole of our planet has been explored, when tribalism is dissipating, when great transnational groupings of states are being organized, when stunning technological advances in communications and transportation are eroding the cultural differences among the various segments of mankind.

But cultural diversity is the forge for the survival of our civilization, just as biological diversity is the forge for the survival of life.

The Earth is overcrowded. Not yet in a literal sense: Our technology is adequate to maintain comfortably a population significantly larger than our present 3.6 billion. The Earth is overcrowded in a psychological sense. For that restless and ambition-driven fraction of mankind that has blazed new paths for our species, there are no new places to go. There are places inside of ourselves, but this is not the forte of such individuals. There are the ocean basins, but we are not yet committed to exploring them seriously; and when we do, they are likely to be exploited rapidly.

At just this time in our history comes the possibility of exploring and colonizing our neighboring worlds in space. The opportunity has come to us not a moment too soon.

October 12, 1992, will be the five hundredth anniversary of the discovery of the "New World" by Christopher Columbus. Mankind will be, I think, engaged at that very moment in an enterprise similar to Columbus'. We will have advantages over him and the mariners of his time. We know precisely where we are going and how to get there. The way will have been examined by unmanned vessels going before us. The paths will have been charted exactly. There will be hazards—collisions with asteroids on voyages to the outer Solar System, for example, or mechanical failure. But there will be no fear of slipping off the edge of the world, as many of the sailors of Columbus' time truly feared. And very likely there will not be a Solar System equivalent of doldrums or sea monsters. Yet the same thrill of exploration and the same adventuresome spirit that drove Columbus will be driving us. As the discovery and exploration of the New World had a profound and irreversible effect on European civilization, exploration and colonization of the Solar System will produce permanent changes in the history and development of mankind.

The analogy with the epic sea voyages of centuries ago is, it seems to me, remarkably close. There was the initial set of sea voyages by Columbus, an Italian in the Spanish court. Our initial set of manned Apollo explorations of the Moon was motivated in significant part by a group of expatriate German engineers led by Wernher von Braun. After Columbus' four voyages, there was essentially a hiatus of a decade or so—and then a burst of further exploratory activities by the Spanish, English, French, and Dutch—vessels flying many flags, many expeditions organized by foreign nationals.

Apollo 17 marked the end of the Apollo lunar missions. It seems clear, at least in the United States, that there will be a hiatus of a decade or more before further lunar exploration and lunar bases are organized. Apollo's primary orientation was never scientific. It was conceived at a time of political embarrassment for the United States. Several historians have suggested that a principal motivation of President Kennedy in organizing the Apollo program was to deflect public attention from the stinging defeat suffered at the Bay of Pigs invasion. Several tens of billions of dollars have been expended on the Apollo program. If the objective had been scientific exploration of the Moon, it could have been carried out much more effectively, for much less money, by unmanned vehicles. The early Apollo missions went to lunar sites of little scientific interest, because the safety of the astronauts was the prime, almost the only, concern. Only toward the very end of the Apollo series did scientific considerations play a significant role.

The Apollo program ended just as the first scientist landed on the Moon. Harrison "Jack" Schmitt, a geologist, trained at Harvard, was one of the two-man crew of the *Apollo 17* landing module. He was the first scientist to study the Moon from the surface of the Moon. It is ironic that just as the Apollo program became able to achieve this major advance in the scientific exploration of the Moon, it was canceled. Fittingly enough, the first scientist to land on the Moon was the last man to land on the Moon—at least in the foreseeable future. There are no plans for follow-on manned missions to the

Moon either by the United States or, so far as we know, by the Soviet Union.

The argument for cancellation of Apollo was economic. Yet the incremental cost of a given mission was in the many tens of millions of dollars, something like one thousandth the total cost of the Apollo program. It is very much as if, against the advice of my wife, I purchase a Rolls-Royce automobile. She argues that a Volkswagen could get me round just as well, but I feel that a Rolls-Royce would take my mind off the troubles of my job. I then spend so much money on the Rolls-Royce that, after driving it a little bit, I find I can drive it no more because I cannot afford the price of a tank of gas—which is about one thousandth the cost of a Rolls-Royce.

I was one of the scientists opposed to an early Apollo mission. But once the Apollo technology was in hand, I was very much for its continuing usage. I believe the wrong decision was made twice—once in opting for early manned missions to the Moon, and later in abandoning such missions. After *Apollo 17*, the United States is left with no program, manned or unmanned, for exploration of the Moon. The Soviet Union has developed, in its Luna series of unmanned spacecraft, a proven and versatile capability for roving exploration of the lunar surface and automatic sample return to Earth.

The example of the earliest exploration of the New World suggests that the hiatus in space will be only temporary. The linkage of Soyuz and Skylab, the orbital stations of the Soviet Union and the United States, scheduled for 1975 or 1976, is presaged as the predecessor for joint manned planetary missions.

The Solar System is much vaster than the Earth, but the speeds of our spacecraft are, of course, much greater than the speeds of the sailing ships of the fifteenth and sixteenth centuries. The spacecraft trip from the Earth to the Moon is faster than was the galleon trip from Spain to the Canary Islands. The voyage from Earth to Mars will take as long as did the sailing time from England to North America; the journey from Earth to the moons of Jupiter will require about the same time as did the voyage from France to Siam in the eighteenth century. Moreover, the fraction of the gross national product of the United States or the Soviet Union that is being expended even in the more costly manned space programs is just comparable to the fraction of the gross national product spent by England and France in the sixteenth and seventeenth centuries on their exploratory ventures by sailing ships. In economic terms and in human terms, we have performed such voyages before!

I believe we will see semipermanent bases on the Moon by the 1980s. They will initially be resupplied with material and personnel from Earth, but will become increasingly self-sustaining, utilizing lunar resources. There will be children born in such colonies. They will eventually think of the Earth as "the old country"—an old-fashioned world in many senses, set in its ways, not moving with the times, more constrained and less free than the lunar colonies, despite the rigors and technological constraints of life on the Moon.

In the comparatively near future the entire Solar System will be explored by sophisticated unmanned space vehicles. I think we will see by the 1980s and 1990s deep-entry probes into the atmospheres of Jupiter and Saturn and Titan (the biggest moon of Saturn)—places that are, I believe, far and away the most favorable in the Solar System for indigenous life; we will witness passages of small spacecraft through comets, landings on the large satellites of Jupiter and Saturn, flybys as far as Neptune and Pluto, and hardy spacecraft that plunge into the Sun, radioing back data before they sear and melt in the interior inferno of the nearest star.

Human landings on even the nearer planets, however, will not be as easy as had once been thought. The surface of Venus, far from being Eden, turns out, as we have seen, to be far more like Hell. We cannot imagine manned exploration of the Venus surface in the next few decades. Venus is a planet with fiery temperatures, noxious gases, and crushing atmospheric pressures. Yet, the clouds of Venus are in a clement environment; and a manned buoyant probe—something like a nineteenth-century balloon gondola in which the astronauts work in shirtsleeves and leather oxygen masks—is not without its charm or its scientific interest.

Mars is a vastly exciting planet, of enormous geological, meteorological, and biological interest. A manned expedition to Mars would be very desirable, except for two objections. First, the cost would be crushing. One hundred billion to two hundred billion dollars is probably a conservative estimate. I cannot bring myself to believe that such an expenditure is necessary in the next few decades—when there is so much misery on Earth that could be relieved by such expenditures. Yet in the longer term, say, in the first decades of the twenty-first century, I do not think that such cost objections will be cogent—particularly because new propulsion and life-support systems will be developed.

The second objection to manned missions to Mars is more subtle. It is equally an objection to automatically returned samples from Mars, like the Soviet Union's Luna series for automatic sample return from the Moon. This is the danger of "back contamination." Precisely because Mars is an environment of great potential biological interest, it is possible that on Mars there are pathogens, organisms which, if transported to the terrestrial environment, might do enormous biological damage—a Martian plague, the twist in the plot of H. G. Wells' *War of the Worlds*, but in reverse. This is an extremely grave point. On the one hand, we can argue that Martian organisms cannot cause any serious problems to terrestrial organisms, because there has been no biological contact for 4.5 billion years between Martian and terrestrial organisms. On the other hand, we can argue equally well that terrestrial organisms have evolved no defenses against potential Martian pathogens, precisely because there has been no such contact for 4.5 billion years. The chance of such an infection may be very small, but the hazards, if it occurs, are certainly very high. Wholesale exterminations of native populations in Santo Domingo and Samoa and Tahiti occurred during the early days of sailing-ship exploration for just such reasons. Among the gifts carried by Columbus to the New World was smallpox.

It is no use arguing that samples can be brought back safely to Earth, or to a base on the Moon, and thereby not be exposed to Earth. The lunar base will be shuttling passengers back and forth to Earth; so will a large Earth orbital station. The one clear lesson that emerged from our experience in attempting to isolate Apollo-returned lunar samples is that mission controllers are unwilling to risk the certain discomfort of an astronaut—never mind his death—against the remote possibility of a global pandemic. When *Apollo 11*, the first successful manned lunar-lander, returned to Earth—it was a spaceworthy, but not a very seaworthy, vessel—the agreed-upon quarantine protocol was immediately breached. It was adjudged better to open the *Apollo 11* hatch to the air of the Pacific Ocean and, for all we then knew, expose the Earth to lunar pathogens, than to risk three seasick astronauts. So little concern was paid to quarantine that the aircraft-carrier crane scheduled to lift the command module unopened out of the Pacific was discovered at the last moment to be unsafe. Exit from *Apollo 11* was required in the open sea.

There is also the vexing question of the latency period. If we expose terrestrial organisms to Martian pathogens, how long must we wait before we can be convinced that the pathogen-host relationship is understood? For example, the latency period for leprosy is more than a decade. Because of the danger of back-contamination of Earth, I firmly believe that manned landings on Mars should be postponed until the beginning of the next century, after a vigorous program of unmanned Martian exobiology and terrestrial epidemiology.

I reach this conclusion reluctantly. I, myself, would love to be involved in the first manned expedition to Mars. But an exhaustive program of unmanned biological exploration of Mars is necessary first. The likelihood that such pathogens exist is probably small, but we cannot take even a small risk with a billion lives. Nevertheless, I believe that people will be treading the Martian surface near the beginning of the twenty-first century.

Beyond that, it is just possible to glimpse the outline of

further exploration and colonization. The large moons of Jupiter, and Titan, the biggest moon of Saturn, are major worlds in their own right. Titan is known to have an atmosphere much thicker than that of Mars. These five moons all have large quantities of ices on their surfaces. To make these worlds more habitable, their ices can be tapped for fuel, for the production of food and for the generation of atmospheres. Schemes for similarly terraforming Mars and Venus have been suggested and will very likely be refined and implemented (see Chapter 22). More exotic possibilities, requiring much more advanced technologies, are not at all beyond the likely capabilities of mankind in the next century or two. These include the establishment of bases on and in the asteroids and the short-period comets.

In another century or so, novel forms of propulsion within the Solar System will be developed. One of the most charming of these is solar sailing, the use of the pressure of sunlight and of the protons and electrons in the solar wind for tripping through the Solar System. Enormous sails will be required for such an enterprise, but they can be extremely thin. We can imagine a spacecraft surrounded with tens of miles of golden gossamer-thin sails, delicately and exquisitely furled to catch the solar wind. Remote scientific stations will be on the lookout for the gusts produced by solar flares. Going outward from the Sun will be easy; tacking inward will be more difficult. Spacecraft using solar electric power and nuclear fusion will very likely also be developed in the next century.

In something like two or three centuries, assuming even a modest growth in our technological capability, I would imagine the entire Solar System will have been explored thoroughly —at least to the extent that the Earth has been explored at the present time, some two or three centuries after the first large-scale exploration and colonization activities were begun by European sailing vessels. After that, it is not out of the question that a more wholesale rearrangement of our Solar System will begin, first slowly and then at a more rapid rate —astroengineering projects to move the planets about, to rearrange their masses for the convenience of mankind, his descendants, and his inventions.

By then—perhaps long before then—we may have made contact with other advanced civilizations in the Galaxy. Or perhaps not. In any case, we will be ready in a few centuries for the next step. At just the point the Solar System begins to be filled in—again psychologically, rather than physically—we will be ready for interstellar voyages. That is a limitless prospect, one that can occupy the best exploratory instincts of mankind forever.

Pioneer 10 is the first interstellar spacecraft launched by mankind. It was also the fastest spacecraft launched, to the date of its departure. But it will take eighty thousand years for *Pioneer 10* to reach the distance of the nearest star. Because space is so empty, it will never enter another Solar System. The little golden message aboard *Pioneer 10* will be read, but only if there are interstellar voyagers able to detect and intercept *Pioneer 10*.

I believe that such an interception may occur, but by interstellar voyagers from the planet Earth, overtaking and heaving to this ancient space derelict—as if the *Nina*, with its crew jabbering in Castilian about falling off the edge of the world, were to be intercepted, somewhere off Tristan da Cunha, by the aircraft carrier *John F. Kennedy*.

Human figure and star field. A drawing by Robert Macintyre.

Part Three

BEYOND THE SOLAR SYSTEM

To dance beneath the diamond sky
With one hand wavin' free . . .

—Bob Dylan, *Mr. Tambourine Man*

Three dolphins, conversing with the author. Photograph by the author.

24.

Some of My Best Friends Are Dolphins

The first scientific conference on the subject of communication with extraterrestrial intelligence was a small affair sponsored by the U. S. National Academy of Sciences in Green Bank, West Virginia. It was held in 1961, a year after Project Ozma, the first (unsuccessful) attempt to listen to possible radio signals from civilizations on planets of other stars. Subsequently, there were two such meetings held in the Soviet Union sponsored by the Soviet Academy of Sciences. Then, in September 1971, a joint Soviet-American conference on Communication with Extraterrestrial Intelligence was held near Byurakan, in Soviet Armenia (see Chapter 27). The possibility of communication with extraterrestrial intelligence is now at least semirespectable, but in 1961 it took a great deal of courage to organize such a meeting. Considerable credit should go to Dr. Otto Struve, then director of the National Radio Astronomy Observatory, who organized and hosted the Green Bank meeting.

Among the invitees to the meeting was Dr. John Lilly, then of the Communication Research Institute, in Coral Gables, Florida. Lilly was there because of his work on dolphin intelligence and, in particular, his efforts to communicate with dolphins. There was a feeling that this effort to communicate with dolphins—the dolphin is probably another intelligent species on our own planet—was in some sense comparable to the task that will face us in communicating with an intelligent species on another planet, should interstellar radio communication be established. I think it will be much easier to understand interstellar messages, if we ever pick them up, than dolphin messages (see Chapter 29), if there are any.

The conjectured connection between dolphins and space was dramatized for me much later, at the lagoon outside the Vertical Assembly Building at Cape Kennedy, as I awaited the *Apollo 17* liftoff. A dolphin quietly swam about, breaking water now and again, surveying the illuminated Saturn booster poised for its journey into space. Just checking us all out, perhaps?

Many of the participants at the Green Bank meeting already knew one another. But Lilly was, for many of us, a new quantity. His dolphins were fascinating, and the prospect of possible communication with them was enchanting. (The meeting was made further memorable by the announcement in Stockholm during its progress that one of our participants, Melvin Calvin, had been awarded the Nobel Prize in Chemistry.)

For many reasons, we wished to commemorate the meeting and maintain some loose coherence as a group. Captured as we were by Lilly's tales of the dolphins, we christened ourselves "The Order of the Dolphins." Calvin had a tie pin struck as an emblem of membership; it was a reproduction from the Boston Museum of an old Greek coin showing a boy on a dolphin. I served as a kind of informal co-ordinator of correspondence the few times that there was any "Dolphin" business, all of it the election of new members. In the following year or two, we elected a few others to membership—among them I. S. Shklovskii, Freeman Dyson, and J. B. S. Haldane. Haldane wrote me that membership in an organization that had no dues, no meetings, and no responsibilities was the sort of organization he appreciated; he promised to try hard to live up to the duties of membership.

The Order of the Dolphins is now moribund. It has been replaced by a number of activities on an international scale. But for me the Order of the Dolphins had a special significance—it provided an opportunity to meet with, talk with, and, to some extent, befriend dolphins.

It was my practice to spend a week or two each winter in the Caribbean, mostly snorkeling and scuba diving—examining the nonmammalian inhabitants of the Caribbean waters. Because of my acquaintance and later friendship with John Lilly, I was also able to spend some days with Lilly's dolphins in Coral Gables and in his research station at St. Thomas in the U. S. Virgin Islands.

His institute, now defunct, unquestionably did some good work on the dolphin, including the production of an important atlas of the dolphin brain. While I will be critical here about some of the scientific aspects of Lilly's work, I want to express my admiration for any serious attempt to investigate dolphins and for Lilly's pioneering efforts in particular. Lilly has since moved on to investigations of the human mind from the inside—consciousness expansion, both pharmacologically and nonpharmacologically induced.

I first met Elvar in the winter of 1963. Laboratory research on dolphins had been limited by these mammals' sensitive skin; it was only the development of plastic polymer tanks that permitted long-term residence of dolphins in the laboratory. I was surprised to find that the Communication Research Institute resided in what used to be a bank, and I had visions of a polystyrene tank in each teller's cage, with dolphins counting the money. Before introducing me to Elvar, Lilly insisted that I don a plastic raincoat, despite my assurances that this was entirely unnecessary. We entered a medium-sized room at the far corner of which was a large polyethylene tank. I could immediately see Elvar with his head thrown back out of the water so that the visual fields of each eye overlapped, giving him binocular vision. He swam slowly to the near side of the tank. John, the perfect host, said, "Carl, this is Elvar; Elvar, this is Carl." Elvar promptly slapped his head forward, down onto the water, producing a needle-beam spray of water that hit me directly on the forehead. I had needed a raincoat after all. John said, "Well, I see you two are getting to know each other"—and promptly left.

I was ignorant of the amenities in dolphin/human social interactions. I approached the tank as casually as I could manage and murmured something like "Hi, Elvar." Elvar promptly turned on his back, exposing his abraded, gun-metal-gray belly. It was so much like a dog wanting to be scratched that, rather gingerly, I massaged his underside. He liked it—or at least I thought he liked it. Bottle-nose dolphins have a sort of permanent smile set into their heads.

After a little while, Elvar swam to the opposite side of the tank and then returned, again presenting himself supine—but this time about six inches subsurface. He obviously wanted his belly scratched some more. This was slightly awkward for me because I was outfitted under my raincoat with a full armory of shirt, tie, and jacket. Not wishing to be impolite, I took off my raincoat, removed my jacket, slid my sleeves up onto my wrists without unbuttoning my shirt cuffs, and put my raincoat on again, all the while assuring Elvar I would return momentarily—which I did, finally scratching him six inches below the water. Again he seemed to like it; again, after a few moments, he retreated to the far side of the tank, and then returned. This time he was about a foot subsurface.

My mood of cordiality was eroding rapidly, but it seemed to me that Elvar and I were at least engaged in communication of a kind. So I once again removed my raincoat, rolled up my sleeves, put my raincoat on again, and attended to Elvar. The next sequence found Elvar three or four feet subsurface, awaiting my massage. I could just have reached him were I prepared to discard raincoat and shirt altogether. This, I decided, was going too far. So we gazed at each other for a while in something of a standoff—man and dolphin, with a meter of water between us. Suddenly, Elvar came booming out of the water head first, until only his tail flukes were in contact with the water. He towered over me, doing a kind of slow back-pedaling, then uttered a noise. It was a single "syllable," high-pitched and squeaky. It had, well, a sort of Donald Duck timbre. It sounded to me that Elvar had said "More!"

I bounded out of the room, found John attending to some electronic equipment, and announced excitedly that Elvar had apparently just said "More!"

John was laconic. "Was it in context?" was all he asked.

"Yes, it was in context."

"Good, that's one of the words he knows."

Eventually, John believed that Elvar had learned some dozens of words of English. To the best of my knowledge, no human has ever learned a single word of delphinese. Perhaps this calibrates the relative intelligence of the two species.

Since the time of Pliny, human history has been full of tales of a strange kindred relation between humans and dolphins. There are innumerable authenticated accounts of dolphins saving human beings who would otherwise have drowned, and of dolphins protecting human beings from attack by other sea predators. As recently as September 1972, according to an account in the New York *Times*, two dolphins protected a twenty-three-year-old shipwrecked woman from predatory sharks during a twenty-five-mile swim in the Indian Ocean. Dolphins are pervasive and dominant motifs in the art of some of the most ancient Mediterranean civilizations, including the Nabatean and Minoan. The Greek coin that Melvin Calvin had duplicated for us is an expression of this long-standing relation.

What humans like about dolphins is clear. They are friendly, and faithful; at times they provide us with food (some dolphins have herded sea animals to fishermen); and they occasionally save our lives. Why dolphins should be attracted to human beings, what we do for them, is far less clear. I will propose later in this chapter that what we provide for dolphins is intellectual stimulation and audio entertainment.

John was replete with dolphin anecdotes of first- or second-hand. I remember three stories in particular. In one, a dolphin was captured in the open sea, put aboard a small ship in a plastic tank, and confronted his captors with a set of sounds, whistles, screeches, and drones that had a remarkably imitative character. They sounded like seagulls, fog horns, train whistles—the noises of shore. The dolphin had been captured by shore creatures and was attempting to make shore talk, as a well-brought-up guest would.

Dolphins produce most of their sounds with their blow hole, which produces the spout of water in their cousins the whales, of whom they are close, miniature anatomical copies.

In another tale, a dolphin held in captivity for some time was let loose in the open sea and followed. When it made contact with a school of dolphins, there was an extremely long and involved sequence of sounds from the liberated prisoner. Was it an account of his imprisonment?

Besides their echo-location clicks—a very effective underwater sonar system—dolphins have a kind of whistle, a kind of squeaky-door noise, and the noise made when imitating human speech, as in Elvar's "More!" They are capable of producing quite pure tones, and pairs of dolphins have been known to produce tones of the same frequency and different phase, so that the "beat" phenomenon of wave physics occurs. The beat phenomenon is a lot of fun. If humans could sing pure tones, I am sure we would go on beating for hours.

There is little doubt that the whistle noises are used for dolphin communications. I heard what seemed to be (I may be anthropomorphizing) very plaintive whistles on St. Thomas from a male adolescent dolphin named Peter, who, for a while, was kept in isolation from two adolescent female dolphins. They all whistled a lot at each other. When the three were reunited in the same pool, their sexual activity was prodigious, and they did not whistle much.

Most of the communication among dolphins that I have heard is of the squeaky-door variety. Dolphins seem to be attracted to humans who make similar noises. In March 1971, for example, in a dolphin pool in Hawaii, I spent forty-five minutes of vigorous squeaky-door "conversation" with several dolphins, to at least some of whom I *seemed* to be saying something of interest. In delphinese it may have been stupefying in its idiocy, but it held their attention.

In another story, John told how it was his practice with dolphins of adolescent age and sexual proclivities to separate male and female over the weekend when there would be no experiments. Otherwise, they would do what John, with some delicacy, described as "going on a honeymoon"—which, however desirable to the dolphins, would leave them in no condition for experimentation on Monday morning. In one case, dolphins could pass across a large tank, from one half to the other, only through a heavy, vertically sliding door. One Monday morning John found the door in place but the two dolphins of opposite sex, Elvar and Chi-Chi, on the same side of the barrier. They had gone on a honeymoon. John's experimental protocol would have to wait, and he was angry. Who had forgotten to separate the dolphins on Friday afternoon? But everyone remembered that the dolphins had been separated and the door properly closed.

As a test, the experimenters repeated the conditions. Elvar and Chi-Chi were separated and the heavy door put in place amid Friday-afternoon ceremonies of loud goodbyes, slammings of building doors, and the heavy trodding of exiting feet. But the dolphins were being observed covertly. When all was quiet, they met at the barrier and exchanged a few low-frequency creaking-door noises. Elvar then pushed the door upward at one corner from his side until it wedged; Chi-Chi, from her side, pushed the opposite corner. Slowly, they worked the door up. Elvar came swimming through and was received by the embraces ("enfinments" is not the right word, either) of his mate. Then, according to John's story, those who lay in waiting announced their presence by whistling, hooting, and booing—whereupon, with some appearance of embarrassment, Elvar swam to his half of the pool and the two dolphins worked down the vertical door from their opposite sides.

This story has such an appealing human character to it—even down to a little dollop of Victorian sexual guilt—that I find it unlikely. But there are many things that are unlikely about dolphins.

I am probably one of the few people who has been "propositioned" by a dolphin. The story requires a little background. I went to St. Thomas one winter to dive and to visit Lilly's dolphin station, which was then headed by Gregory Bateson, an Englishman of remarkable and diverse interests in anthropology, psychology, and human and animal behavior. Dining with some friends at a fairly remote mountaintop restaurant, we engaged in casual conversation with the hostess at the restaurant, a young woman named Margaret. She described to me how uneventful and uninteresting her days were (she was hostess only at night). Earlier the same day Bateson had described to me his difficulties in finding adequate research assistants for his dolphin program. It was not difficult to introduce Margaret and Gregory to each other. Margaret was soon working with dolphins.

After Bateson left St. Thomas, Margaret was for a while *de facto* director of the research station. In the course of her work, Margaret performed a remarkable experiment, described in some detail in Lilly's book *The Mind of the Dolphin.* She began living on a kind of suspended raft over the pool of Peter the dolphin, spending twenty-four hours a day in close contact with him. Margaret's experiment occurred not long before the incident I now speak of; it may have had something to do with Peter's attitude toward me.

I was swimming in a large indoor pool with Peter. When I threw the pool's rubber ball to Peter (as was natural for me to have done), he dove under the ball as it hit the water and batted it with his snout accurately into my hands. After a few throws and precision returns, Peter's returns became increasingly inaccurate—forcing me to swim first to one side of the pool and then to the other in order to retrieve the ball. Eventually, it became clear that Peter chose not to place the ball within ten feet of me. He had changed the rules of the game.

Peter was performing a psychological experiment on me—to learn to what extreme lengths I would go to continue this pointless game of catch. It was the same kind of psychological testing that Elvar had conducted in our first meeting. Such testing is one clue to the bond that draws dolphins to humans: We are one of the few species that have pretensions of psychological knowledge; therefore, we are one of the few that would permit, however inadvertently, dolphins to perform psychological experiments on us.

As in my first interview with Elvar, I eventually saw what was happening and decided stoutly that no dolphin was going to perform a psychological experiment on me. So I held the ball and merely tread water. After a minute or so, Peter swam rapidly toward me and made a grazing collision. He circled around and repeated this strange performance. This time I felt some protrusion of Peter's lightly brushing my side as he passed. As he circled for a third pass, I idly wondered what this protrusion might be. It was not his tail flukes, it was not . . . Suddenly it dawned on me, and I felt like some maiden aunt to whom an improper proposal had just been put. I was not prepared to cooperate, and all sorts of conventional expressions came unbidden to my mind—like, "Don't you know any nice girl dolphins?" But Peter remained cheerful and unoffended by my unresponsiveness. (Is it possible, I now wonder, that he thought I was too dense to understand even *that* message?)

Peter had been separated from female dolphins for some time and, in the not too distant past, had spent many days in close contact, including sexual contact, with Margaret, another human being. I do not think that there is any sexual bond that accounts for the closeness that dolphins feel toward humans, but the incident had some significance. Even in what we piously describe as "bestiality" there are only a few species which, so far as I have heard, are put upon by human beings for interspecific sexual activities; these are entirely of the sort that humans have domesticated. I wonder if some dolphins have thoughts about domesticating us.

Dolphin anecdotes make marvelous cocktail party accounts, an unending source of casual conversation. One of the difficulties that I discovered with research into dolphin language and intelligence was precisely this fascination with anecdote; the really critical scientific tests were somehow never performed.

For example, I repeatedly urged that the following experiment be done: Dolphin A is introduced into a tank that is equipped with two underwater audio speakers. Each hydrophone is attached to an automatic fish dispenser catering tasty dolphin fare. One speaker plays Bach, the other plays Beatles.

Which speaker is playing Bach or Beatles (a different composition each time) at any given moment is determined randomly. Whenever Dolphin A goes to the appropriate speaker—let us say, the one playing Beatles—he is rewarded with a fish. I think there is no doubt that any dolphin will—because of his great interest in, and facility with, the audio spectrum—be able soon to distinguish between Bach and Beatles. But that is not the significant part of the experiment. What is significant is the number of trials before Dolphin A becomes sophisticated—that is, always knows that if he wishes a fish he should go to the speaker playing Beatles.

Now Dolphin A is separated from the speakers by a barrier of plastic broad-gauge mesh. He can see through the barrier, he can smell and taste through it, and, most important, he can hear and "speak" through it. But he cannot swim through it. Dolphin B is then introduced into the area of the speakers. Dolphin B is naïve; that is, he has had no prior experience with underwater fish dispensers, Bach, or Beatles. Unlike the well-known difficulty in finding "naïve" college students with whom to perform experiments on *cannibis sativa,* there should be no difficulty finding dolphins lacking extensive experience with Bach and Beatles. Dolphin B must go through the same learning procedure as did Dolphin A. But now each time that Dolphin B (at first randomly) succeeds, not only does the dispenser provide him with a fish, but a fish is also thrown to Dolphin A, who is able to witness the learning experience of Dolphin B. If Dolphin A is hungry, it is distinctly to his advantage to communicate what he knows about Bach and Beatles to Dolphin B. If Dolphin B is hungry, it is to his advantage to pay attention to the information that Dolphin A may have. The question, therefore, is: Does Dolphin B have a steeper learning curve than Dolphin A? Does he reach the plateau of sophistication in fewer trials or less time?

If such experiments were repeated many times and it were found that the learning curves for Dolphin B were in a statistically significant sense always steeper than those of Dolphin A, communication of moderately interesting information between two dolphins would have been established. It might be a verbal description of the difference between Bach and the Beatles—to my mind, a difficult but not impossible task—or it might simply be the distinction between left and right in each trial, until Dolphin B catches on. This is not the best experimental design to test dolphin-to-dolphin communication, but it is typical of a large category of experiments that could be performed. To my knowledge and regret, no such experiments have been performed with dolphins to date.

Questions of dolphin intelligence have taken on a special poignancy for me in the past few years as the case of the humpback whale unfolded. In a remarkable set of experiments, Roger Payne, of Rockefeller University, has trailed hydrophones to a depth of tens of meters in the Caribbean, seeking and recording the songs of the humpback whale. Another member, along with the dolphins, of the taxonomic class of Cetacea, the humpback whale has extraordinarily complex and beautiful articulations, which carry over considerable distances beneath the ocean surface, and which have an apparent social utility within and between schools of whales, which are very gregarious social animals.

The brain size of whales is much larger than that of humans. Their cerebral cortexes are as convoluted. They are at least as social as humans. Anthropologists believe that the development of human intelligence has been critically dependent upon these three factors: Brain volume, brain convolutions, and social interactions among individuals. Here we find a class of animals where the three conditions leading to human intelligence may be exceeded, and in some cases greatly exceeded.

But because whales and dolphins have no hands, tentacles, or other manipulative organs, their intelligence cannot be worked out in technology. What is left? Payne has recorded examples of very long songs sung by the humpback whale; some of the songs were as long as half an hour or more. A few of them appear to be repeatable, virtually phoneme by phoneme; somewhat later the entire cycle of sounds comes out virtually identically once again. Some of the songs have been commercially recorded and are available on CRM Records

(SWR-II). I calculate that the approximate number of bits (see Chapter 34) of information (individual yes/no questions necessary to characterize the song) in a whale song of half an hour's length is between a million and a hundred million bits. Because of the very large frequency variation in these songs, I have assumed that the frequency is important in the content of the song—or, put another way, that whale language is tonal. If it is not as tonal as I guess, the number of bits in such a song may go down by a factor of ten. Now, a million bits is approximately the number of bits in *The Odyssey* or the Icelandic *Eddas*. (It is also unlikely that in the few hydrophone forays into Cetacean vocalizations that have been made to date, the longest of such songs has been recorded.)

Is it possible that the intelligence of Cetaceans is channeled into the equivalent of epic poetry, history, and elaborate codes of social interaction? Are whales and dolphins like human Homers before the invention of writing, telling of great deeds done in years gone by in the depths and far reaches of the sea? Is there a kind of *Moby Dick* in reverse—a tragedy, from the point of view of a whale, of a compulsive and implacable enemy, of unprovoked attacks by strange wooden and metal beasts plying the seas and laden with humans?

The Cetacea hold an important lesson for us. The lesson is not about whales and dolphins, but about ourselves. There is at least moderately convincing evidence that there is another class of intelligent beings on Earth besides ourselves. They have behaved benignly and in many cases affectionately toward us. We have systematically slaughtered them. There is a monstrous and barbaric traffic in the carcasses and vital fluids of whales. Oil is extracted for lipstick, industrial lubricants and other purposes, even though this makes, at best, marginal economic sense—there are effective substitute lubricants. But why, until recently, has there been so little outcry against this slaughter, so little compassion for the whale?

Little reverence for life is evident in the whaling industry—underscoring a deep human failing that is, however, not restricted to whales. In warfare, man against man, it is common for each side to dehumanize the other so that there will be none of the natural misgivings that a human being has at slaughtering another. The Nazis achieved this goal comprehensively by declaring whole peoples *untermenschen,* subhumans. It was then permissible, after such reclassification, to deprive these peoples of civil liberties, enslave them, and murder them. The Nazis are the most monstrous, but not the most recent, example. Many other cases could be quoted. For Americans, covert reclassifications of other peoples as *untermenschen* has been the lubricant of military and economic machinery, from the early wars against the American Indians to our most recent military involvements, where other human beings, military adversaries but inheritors of an ancient culture, are decried as gooks, slopeheads, slanteyes, and so on—a litany of dehumanization—until our soldiers and airmen could feel comfortable at slaughtering them.

Automated warfare and aerial destruction of unseen targets make such dehumanization all the easier. It increases the "efficiency" of warfare because it undercuts our sympathies with our fellow creatures. If we do not see whom we kill, we feel not nearly so bad about murder. And if we can so easily rationalize the slaughter of others of our own species, how much more difficult will it be to have a reverence for intelligent individuals of different species?

It is at this point that the ultimate significance of dolphins in the search for extraterrestrial intelligence emerges. It is not a question of whether we are emotionally prepared in the long run to confront a message from the stars. It is whether we can develop a sense that beings with quite different evolutionary histories, beings who may look far different from us, even "monstrous," may, nevertheless, be worthy of friendship and reverence, brotherhood and trust. We have far to go; while there is every sign that the human community is moving in this direction, the question is, are we moving fast enough? The most likely contact with extraterrestrial intelligence is with a society far more advanced than we (Chapter 31). But we will not at any time in the foreseeable future be in the position of the American Indians or the Vietnamese—colonial barbarity practiced on us by a technologically more advanced civilization—because of the great spaces between the stars and what I believe is the neutrality or benignness of any civilization that has survived long enough for us to make contact with it. Nor will the situation be the other way around, terrestrial predation on extraterrestrial civilizations—they are too far away from us and we are relatively powerless. Contact with another intelligent species on a planet of some other star—a species biologically far more different from us than dolphins or whales—may help us to cast off our baggage of accumulated jingoisms, from nationalism to human chauvinism. Though the search for extraterrestrial intelligence may take a very long time, we could not do better than to start with a program of rehumanization by making friends with the whales and the dolphins.

25.

"Hello, Central Casting? Send Me Twenty Extraterrestrials"

My friend Arthur C. Clarke had a problem. He was writing a major motion picture with Stanley Kubrick of *Dr. Strangelove* fame. It was to be called *Journey Beyond the Stars,* and a small crisis in the story development had arisen. Could I have dinner with them at Kubrick's New York penthouse and help adjudicate the dispute? (The film's title, by the way, seemed a little strange to me. As far as I knew, there is no place beyond the stars. A film about such a place would have to be two hours of blank screen—a possible plot only for Andy Warhol. I was sure that was not what Kubrick and Clarke had in mind.)

After a pleasant dinner, the crisis emerged as follows: About midway through the movie, a manned space vehicle is making a close approach to either Jupiter 5, the innermost satellite of Jupiter, or to Iapetus, one of the middle-sized satellites of Saturn. As the spacecraft approaches and the curvature of the satellite is visible on the screen, we become aware that the satellite is not a natural moon. It is an artifact of some immensely powerful, advanced civilization. Suddenly, an aperture appears in the side of the satellite; through it we see—stars. But they are not the stars on the other side of the satellite. They are a portion of the sky from elsewhere. Small drone rockets are fired into the aperture, but contact with them is lost as soon as they pass through. The aperture is a space gate, a way to get from one part of the universe to another without the awkwardness of traversing the intervening distance. The spacecraft plunges through the gate and emerges in the vicinity of another stellar system, with a red giant star blazing in the sky. Orbiting the red giant is a planet, obviously the site of an advanced technological civilization. The spacecraft approaches the planet, makes landfall, and then—what?

Although the human elements were nearing studio production in England, this fairly important plot line—the ending! —had not yet been worked out by the two authors. The spacecraft's crew, or some fraction of it, was to make contact with extraterrestrials. Yes, but how to portray the extraterrestrials? Kubrick favored extraterrestrials not profoundly different from human beings. Kubrick's preference had one distinct advantage, an economic one: He could call up Central Casting and ask for twenty extraterrestrials. With a little makeup, he would have his problem solved. The alternative portrayal of extraterrestrials, whatever it was, was bound to be expensive.

I argued that the number of individually unlikely events in the evolutionary history of Man was so great that nothing like us is ever likely to evolve again anywhere else in the universe. I suggested that any explicit representation of an advanced extraterrestrial being was bound to have at least an element of falseness about it, and that the best solution would be to suggest, rather than explicitly to display, the extraterrestrials.

The film, subsequently titled *2001: A Space Odyssey*, opened three years later. At the premiere, I was pleased to see that I had been of some help. As we learn from Jerome Agel's book *The Making of Kubrick's "2001"* (Signet, 1970), Kubrick experimented during production with many representations of extraterrestrial life, including a pirouetting dancer in black tights with white polka dots. Photographed against a black background, this would have been visually very effective. He finally decided on a surrealistic representation of extraterrestrial intelligence. The movie has played a significant role in expanding the average person's awareness of the cosmic perspective. Many Soviet scientists consider *2001* to be the best American movie they have seen. The extraterrestrial ambiguities did not bother them at all.

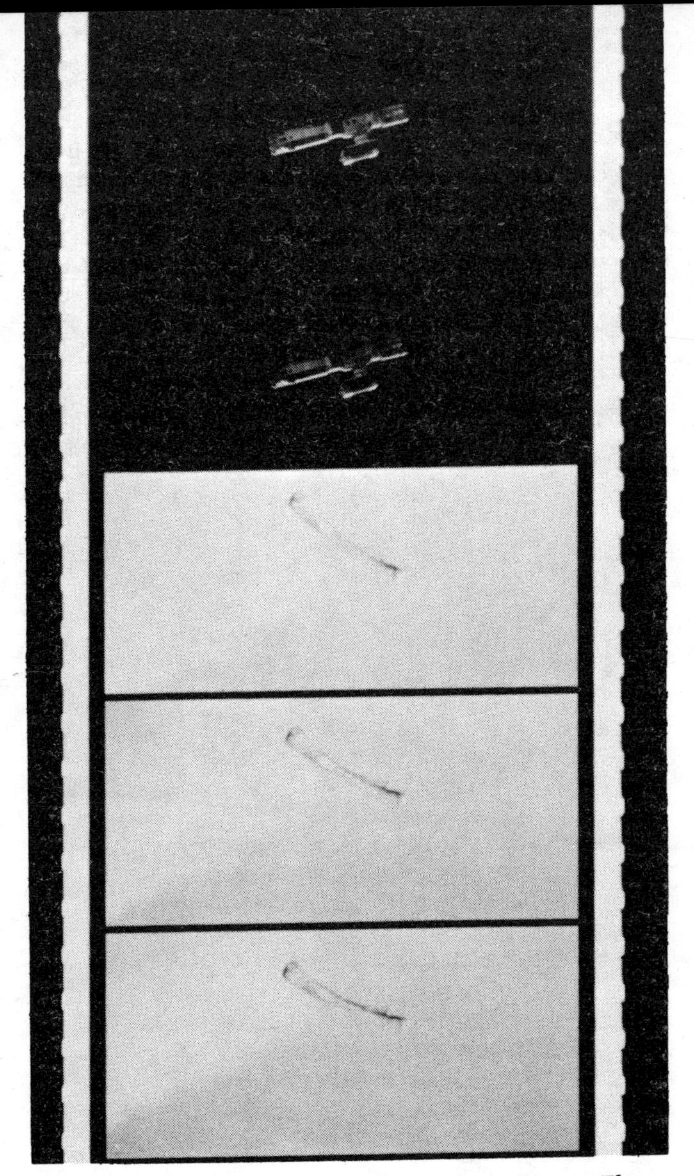

Sequence from *2001: A Space Odyssey*. Cover picture, *The Making of Kubrick's "2001,"* original Signet paperback.

During the filming of *2001,* Kubrick, who obviously has a grasp for detail, became concerned that extraterrestrial intelligence might be discovered before the $10.5 million film was released, rendering the plot line obsolete, if not erroneous. Lloyd's of London was approached to underwrite an insurance policy protecting against losses should extraterrestrial intelligence be discovered. Lloyd's of London, which insures against the most implausible contingencies, declined to write such a policy. In the mid-1960s there was no search being performed for extraterrestrial intelligence, and the chance of accidentally stumbling on extraterrestrial intelligence in a few years' period was extremely small. Lloyd's of London missed a good bet.

26.
The Cosmic Connection

From earliest times, human beings have pondered their place in the universe. They have wondered whether they are in some sense connected with the awesome and immense cosmos in which the Earth is imbedded.

Many thousands of years ago a pseudoscience called astrology was invented. The positions of the planets at the birth of a child were supposed to play a major role in determining his or her future. The planets, moving points of light, were thought, in some mysterious sense, to be gods. In his vanity, Man imagined the universe designed for his benefit and organized for his use.

Perhaps the planets were identified with gods because their motions seemed irregular. The word "planet" is Greek for wanderer. The unpredictable behavior of the gods in many legends may have corresponded well with the apparently unpredictable motions of the planets. The argument may have been: Gods don't follow rules; planets don't follow rules; planets are gods.

When the ancient priestly astrological caste discovered that the motions of the planets were not irregular but predictable, they seem to have kept this information to themselves. No use unnecessarily worrying the populace, undermining religious belief, and eroding the supports of political power. Moreover, the Sun was the source of life. The Moon, through the tides, dominated agriculture—especially in river basins like the Indus, the Nile, the Yangtze, and the Tigris-Euphrates. How reasonable that these lesser lights, the planets, should have a subtler but no less definite influence on human life!

The search for a connection, a hooking-up between people and the universe, has not diminished since the dawn of astrology. The same human needs exist despite the advances of science.

We now know that the planets are worlds more or less like our own. We know that their light and gravity have negligible influence on a newborn babe. We know that there are enormous numbers of other objects—asteroids, comets, pulsars, quasars, exploding galaxies, black holes, and the rest—objects not known to the ancient speculators who invented astrology. The universe is immensely grander than they could have imagined.

Astrology has not attempted to keep pace with the times. Even the calculations of planetary motions and positions performed by most astrologers are usually inaccurate.

No study shows a statistically significant success rate in predicting through their horoscopes the futures or the personality traits of newborn children. There is no field of radio-astrology or X-ray astrology or gamma-ray astrology, taking account of the energetic new astronomical sources discovered in recent years.

Nevertheless, astrology remains immensely popular everywhere. There are at least ten times more astrologers than astronomers. A large number, perhaps a majority, of newspapers in the United States have daily columns on astrology.

Many bright and socially committed young people have more than a passing interest in astrology. It satisfies an almost unspoken need to feel a significance for human beings in a vast and awesome cosmos, to believe that we are in some way hooked up with the universe—an ideal of many drug and religious experiences, the *samadhi* of some Eastern religions.

The great insights of modern astronomy have shown that, in some senses quite different from those imagined by the earlier astrologers, we *are* connected up with the universe.

The first scientists and philosophers—Aristotle, for example—imagined that the heavens were made of a different sort of material than the Earth, a special kind of celestial stuff, pure and undefiled. We now know that this is not the case. Pieces

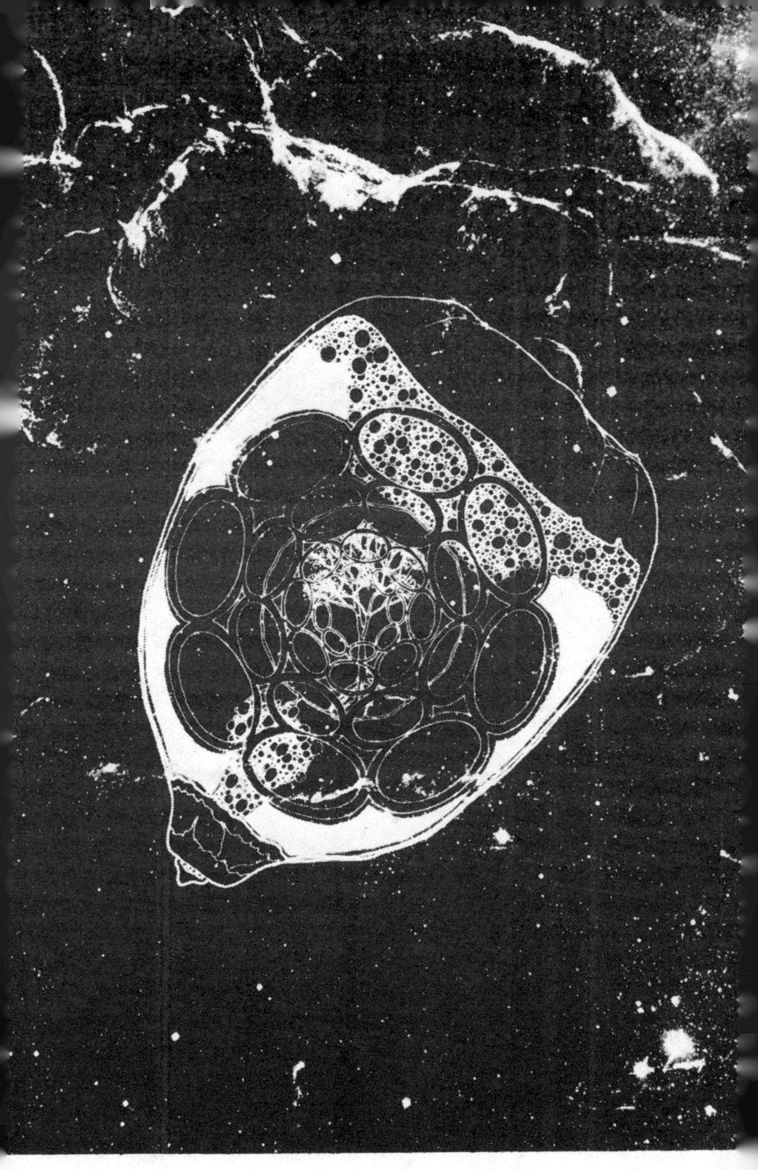

Acorn by Jon Lomberg. The faint wisps of bright gas in the background are from a supernova remnant.

of the asteroid belt called meteorites; samples of the Moon returned by Apollo astronauts and Soviet unmanned spacecraft; the solar wind, which expands outward past our planet from the Sun; and the cosmic rays, which are probably generated from exploding stars and their remnants—all show the presence of the same atoms we know here on Earth. Astronomical spectroscopy is able to determine the chemical composition of collections of stars billions of light-years away. The entire universe is made of familiar stuff. The same atoms and molecules occur at enormous distances from Earth as occur here within our Solar System.

These studies have yielded a remarkable conclusion. Not only is the universe made everywhere of the same atoms, but the atoms, roughly speaking, are present everywhere in approximately the same proportions.

Almost all the stuff of the stars and the interstellar matter between the stars is hydrogen and helium, the two simplest atoms. All other atoms are impurities, trace constituents. This is also true for the massive outer planets of our Solar System, like Jupiter. But it is not true for the comparatively tiny hunks of rock and metal in the inner part of the Solar System, like our planet Earth. This is because the small terrestrial planets have gravities too weak to hold their original hydrogen and helium atmospheres, which have slowly leaked away to space.

The next most abundant atoms in the universe turn out to be oxygen, carbon, nitrogen, and neon. These are atoms everyone has heard of. Why are the cosmically most abundant elements those that are reasonably common on Earth—rather than, say, yttrium or praseodymium?

The theory of the evolution of stars is sufficiently advanced that astronomers are able to understand the various kinds of stars and their relations—how a star is born from the interstellar gas and dust, how it shines and evolves by thermonuclear reactions in its hot interior, and how it dies. These thermonuclear reactions are of the same sort as the reactions that underlie thermonuclear weapons (hydrogen bombs): The conversion of four atoms of hydrogen into one of helium.

But in the later stages of stellar evolution, higher temperatures are reached in the insides of stars, and elements heavier than helium are generated by thermonuclear processes. Nuclear astrophysics indicates that the most abundant atoms produced in such hot red giant stars are precisely the most abundant atoms on Earth and elsewhere in the universe. The heavy atoms generated in the insides of red giants are spewed out into the interstellar medium, by slow leakage from the star's atmosphere like our own solar wind, or by mighty stellar explosions, some of which can make a star a billion times brighter than our Sun.

Recent infrared spectroscopy of hot stars has discovered that they are blowing off silicates into space—rock powder spewed out into the interstellar medium. Carbon stars probably expel graphite particles into surrounding cosmic space. Other stars shed ice. In their early histories, stars like the Sun probably propelled large quantities of organic compounds into interstellar space; indeed, simple organic molecules are found by radio astronomical methods to be filling the space between the stars. The brightest planetary nebula known (a planetary nebula is an expanding cloud usually surrounding an exploding star called a nova) seems to contain particles of magnesium carbonate: Dolomite, the stuff of the European mountains of the same name, expelled by a star into interstellar space.

These heavy atoms—carbon, nitrogen, oxygen, silicon, and the rest—then float about in the interstellar medium until, at some later time, a local gravitational condensation occurs and a new sun and new planets are formed. This second-generation solar system is enriched in heavy elements.

The fate of individual human beings may not now be connected in a deep way with the rest of the universe, but the matter out of which each of us is made is intimately tied to processes that occurred immense intervals of time and enormous distances in space away from us. Our Sun is a second- or third-generation star. All of the rocky and metallic material we stand on, the iron in our blood, the calcium in our teeth, the carbon in our genes were produced billions of years ago in the interior of a red giant star. We are made of star-stuff.

Our atomic and molecular connection with the rest of the universe is a real and unfanciful cosmic hookup. As we explore our surroundings by telescope and space vehicle, other hookups may emerge. There may be a network of intercommunicating extraterrestrial civilizations to which we may link up tomorrow, for all we know. The undelivered promise of astrology—that the stars impel our individual characters—will not be satisfied by modern astronomy. But the deep human need to seek and understand our connection with the universe is a goal well within our grasp.

27.

Extraterrestrial Life: An Idea Whose Time Has Come

Thousands of years ago, the idea that the planets were populated by intelligent beings was uncommon. The idea was that the planets themselves were intelligent beings. Mars was the god of war, Venus was the goddess of beauty, Jupiter was the king of the gods.

In early Roman times a few writers, for example Lucian of Samasota, conceived that at least the Moon was a place that was populated as the Earth was. His science-fiction story describing travel to the Moon was called the "True History." It was, of course, false.

The idea of the planets as an elegant celestial clockwork created by the Deity for the amazement and utility of men emerged in the Renaissance. In the year 1600 Giordano Bruno was burned to death at the stake, in part for uttering and publishing the heresy that there were other worlds and other beings inhabiting them.

The pendulum swung far in the other direction in subsequent centuries. Writers such as Bernard de Fontenelle, Emanuel Swedenborg, and even Immanuel Kant and Johannes Kepler could safely imagine that perhaps all the planets were inhabited. Indeed, the idea was expressed that the name of the planet gave some hint to the character of its inhabitants. The denizens of Venus were amorous; those of Mars, warlike or martial; the inhabitants of Mercury, fickle or mercurial; those of Jupiter, jolly or jovial. And so on. The great British astronomer William Herschel even supposed that the Sun was inhabited.

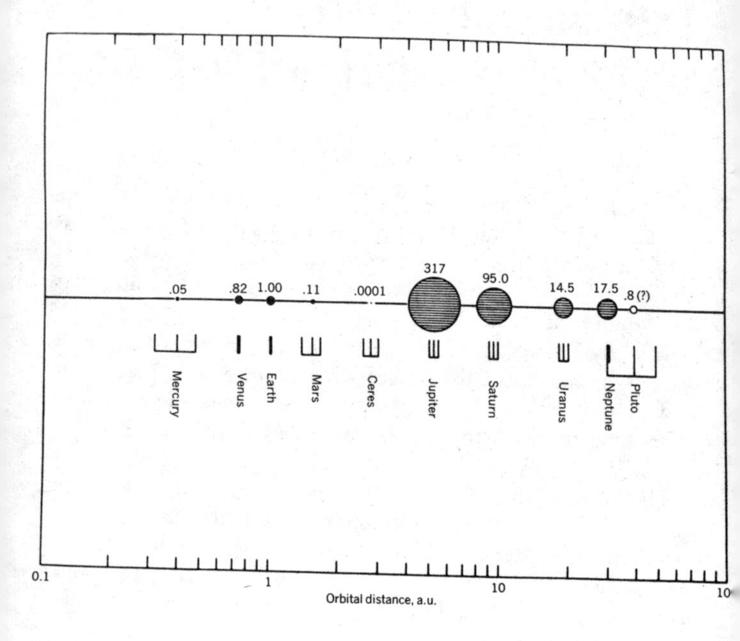

The Solar System. Distances from the Sun are shown in units of the Earth's distance. The trident markings are a sign of the eccentricity of a planetary orbit and show the closest, farthest, and average distance of the planet from the Sun. The masses of the planets are shown in units of the Earth's mass. The Jovian planets are distinguished from the terrestrial planets by cross-hatching.

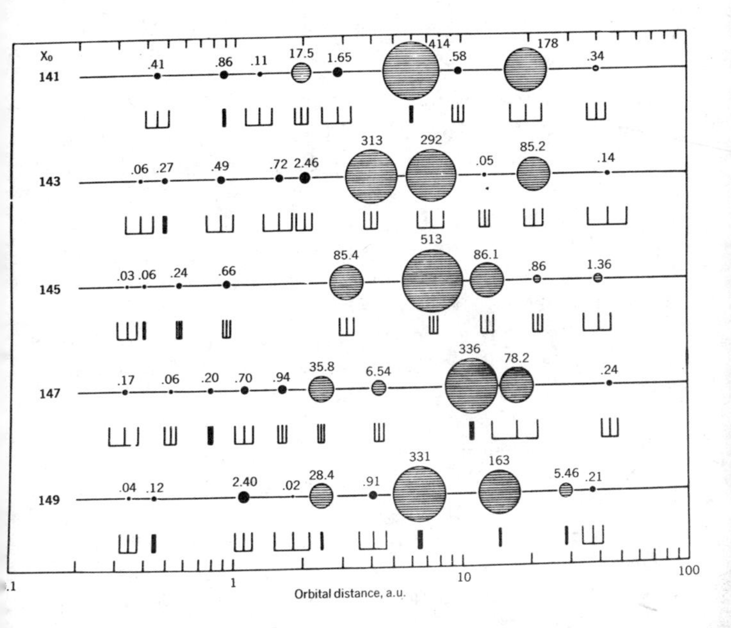

Five model solar systems derived by Steven Dole in a computer experiment on the physics of the origins of solar systems. Dole's systems are clearly very similar to our Solar System. This is one of several lines of evidence suggesting that planetary systems are common accompaniments of stars throughout the Galaxy. Courtesy, ICARUS.

But as the extremes of the physical environments in the Solar System became clearer and the exquisite adaptation to the environment of organisms on Earth became more apparent, skeptics arose. Perhaps Mars and Venus were inhabited, but surely not Mercury, not the Moon, not Jupiter. And so on.

In the last few decades of the nineteenth century the observations of the planet Mars by Giovanni Schiaparelli and Percival Lowell quickened public excitement about the possibility of intelligence on our planetary neighbor. Lowell's passion for the idea of intelligent beings on Mars, his articulateness, and the wide publication of his books did much to bring this idea to the public attention, as did science-fiction writers who followed the Lowellian scenario.

But as the evidence for intelligent life on Mars withered, and as the environment of Mars was perceived to be more and more inclement by terrestrial standards, popular enthusiasm for the idea waned.

By then, scientific interest in extraterrestrial life had reached a nadir. The very enthusiasm with which Lowell pursued the idea of intelligent beings on Mars and the attention that these ideas received from the man in the street repelled many scientists. In addition, a new astronomical field, astrophysics, the application of physics to the surfaces and interiors of stars, had achieved phenomenal success, and the brightest and most enthusiastic young astronomers went into stellar astronomy rather than planetary studies. The pendulum had swung so far that in the period just after the Second World War, there was—in all of the United States—only one astronomer doing serious physical investigations of the planets, G. P. Kuiper, then of the University of Chicago. Not only had astronomers been turned off extraterrestrial life, they had been turned off planetary studies in general.

Since 1950, the situation has slowly reversed again; the pendulum is once more swinging. The development of new measuring instruments (a by-product of World War II), at first ground-based and then, more important, space-borne, has produced a massive infusion of basic new knowledge about the physical environments of the Moon and planets. Young scientists have again been attracted to planetary studies, not only astronomers, but also geologists, chemists, physicists, and biologists. The discipline needs them all.

We now know that the building blocks for the origin of life are in the cards of physics and chemistry; whenever standard primitive atmospheres are exposed to common energy sources, the building blocks of life on Earth drop out of the atmosphere in times of days or weeks. Organic compounds have been found in meteorites and in interstellar space. Small quantities have been found even in such an inhospitable environment as the Moon. They are suspected to exist in Jupiter, in the outer planets of the Solar System, as well as on Titan, the largest moon of Saturn. Both theory and observation now suggest that planets are a common, if not invariable, accompaniment of stars, rather than an exceedingly rare occurrence, as was fashionable to believe in the first decades of this century (see pages 192 and 193).

We now have, for the first time, the tools to make contact with civilizations on planets of other stars. It is an astonishing fact that the great one-thousand-foot-diameter radio telescope of the National Astronomy and Ionosphere Center, run by Cornell University in Arecibo, Puerto Rico, would be able to communicate with an identical copy of itself anywhere in the Milky Way Galaxy. We have at our command the means to communicate not merely over distances of hundreds or thousands of light-years; we can communicate over tens of thousands of light-years, into a volume containing hundreds of billions of stars. The hypothesis that advanced technical civilizations exist on planets of other stars is amenable to experimental testing. It has been removed from the arena of pure speculation. It is now in the arena of experiment.

Our first attempt to listen to broadcasts from extraterrestrial societies was Project Ozma. Organized by Frank Drake in 1960 at the National Radio Astronomy Observatory (NRAO), it looked at two stars at one frequency for two weeks. The results were negative. Slightly more ambitious projects are, at the time of writing, being performed at the Gorky Radiophysical Institute in the Soviet Union and at NRAO in the United States. All in all, perhaps a few hundred nearby stars will be examined at one or two frequencies. But even the most optimistic calculations on the distances to the nearest stars suggest that hundreds of thousands to millions of stars must be examined before an intelligible signal from one of them will be received. This requires a large effort covering a sizable period of time. But it is well within our resources, our abilities, and our interests.

The change in the climate of opinion about extraterrestrial life was reflected in 1971 by a scientific conference held in Byurakan, Soviet Armenia, and sponsored jointly by the Soviet Academy of Sciences of the U.S.S.R. and the National Academy of Sciences of the United States. I had the privilege of chairing the U.S. delegation to this meeting. The participants represented astronomy, physics, mathematics, biology, chemistry, archaeology, anthropology, history, electronics, computer technology, and cryptography. The group, which included two skeptical Nobel laureates, was marked for its crossing of national as well as disciplinary boundaries. The conference concluded that the chances of there being extraterrestrial communicative societies and our present technological ability to contact them were both sufficiently high that a serious search was warranted. Some of the specific conclusions that were reached were these:

1. The striking discoveries of recent years in the fields of astronomy, biology, computer science and radiophysics have transferred some of the problems of extraterrestrial civilizations and their detection from the realm of speculation to a new realm of experiment and observation. For the first time in human history, it has become possible to make serious and detailed experimental investigations of this fundamental and important problem.

2. This problem may prove to be of profound significance for the future development of Mankind. If extraterrestrial civilizations are ever discovered, the affect on human scientific and technological capabilities will be im-

mense, and the discovery can positively influence the whole future of Man. The practical and philosophical significance of a successful contact with an extraterrestrial civilization would be so enormous as to justify the expenditure of substantial efforts. The consequences of such a discovery would greatly add to the total of human knowledge.

3. The technological and scientific resources of our planet are already large enough to permit us to begin investigations directed towards the search for extraterrestrial intelligence. As a rule, such studies should provide important scientific results even when specific searches for extraterrestrial intelligence do not succeed. At present, these investigations can be carried out effectively in the various countries by their own scientific institutions. Even at this early stage, however, it would be useful to discuss and coordinate specific programs of research and to exchange scientific information. In the future, it would be desirable to combine the efforts of investigators in various countries to achieve the experimental and observational objectives. It seems to us appropriate that the search for extraterrestrial intelligence should be made by representatives of the whole of mankind.

4. Various modes of search for extraterrestrial intelligence were discussed in detail at the Conference. The realization of the most elaborate of these proposals would require considerable time and effort and an expenditure of funds comparable to the funds devoted to space and nuclear research. Useful searches can, however, also be initiated at a very modest scale.

5. The Conference participants consider highly valuable present and forthcoming space-vehicle experiments directed towards searching for life on the other planets of our solar system. They recommend the continuation and strengthening of work in such areas as prebiological organic chemistry, searches for extrasolar planetary systems, and evolutionary biology, which bear sharply on the problem.

6. The Conference recommends the initiation of specific new investigations directed towards modes of search for signals.

(The complete Proceedings of the conference are published as *Communication with Extraterrestrial Intelligence*, Carl Sagan, ed., Cambridge, Massachusetts, The M.I.T. Press, 1973.)

Another sign of the increasing acceptability of the search for extraterrestrial intelligence is the recommendations of the Astronomy Survey Committee of the U. S. National Academy of Sciences, which had been asked to summarize the needs of astronomy in the decade of the 1970s. The Committee's report was the first such national report on the future of astronomy to lay stress on the search for extraterrestrial intelligence—as a possibly important by-product of astronomical research in the near future and as a justification for the construction of large radio telescopes.

Nearer to home, there is an accelerating set of laboratory studies of the origin of life on Earth. If the origin of life on Earth turns out to have been exceedingly "easy," the chances of life elsewhere are correspondingly high.

There is also a concerted effort in the United States—Project Viking—to land instrumented payloads on the surface of Mars to search for indigenous life forms.

The idea of extraterrestrial life is an idea whose time has come.

28.
Has the Earth Been Visited?

By far the cheapest way of communicating with the Earth, if you're a representative of an advanced extraterrestrial civilization, is by radio. A single bit of radio information, sent winging across space to the Earth, would cost far less than a penny. A radio search for extraterrestrial intelligence seems, therefore, a very reasonable place for us to begin. But should we not examine other possibilities closer to home? Wouldn't we look silly if we expended a major effort listening for radio messages or searching for life on Mars if, all the while, there was here on Earth evidence of extraterrestrial life?

There are two hypotheses of this sort that have gained a following in the popular literature. The first postulates that the Earth is today being visited by spacecraft from other worlds —this is the extraterrestrial flying saucer or unidentified flying object (UFO) hypothesis. The second also postulates that the Earth has been visited by such spacecraft, but in the past, before written history.

The extraterrestrial hypothesis of UFO origins is a complex subject, powerfully dependent on the reliability of witnesses. A comprehensive discussion of this problem has recently been published in *UFO's: A Scientific Debate* (Carl Sagan and Thornton Page, editors, Ithaca, N.Y., Cornell University Press, 1972), in which all sides of the subject have been aired. My own view is that there are no cases that are simultaneously very reliable (reported independently by a large number of witnesses) and very exotic (not explicable in terms of reasonably postulated phenomena—as a strange moving light could be a searchlight from a weather airplane or a military aerial refueling operation). There are no reliably reported cases of strange machines landing and taking off, for example.

There is another approach to the extraterrestrial hypothesis of UFO origins. This assessment depends on a large number of factors about which we know little, and a few about which we know literally nothing. I want to make some crude numerical estimate of the probability that we are frequently visited by extraterrestrial beings.

Now, there is a range of hypotheses that can be examined in such a way. Let me give a simple example: Consider the Santa Claus hypothesis, which maintains that, in a period of eight hours or so on December 24–25 of each year, an outsized elf visits one hundred million homes in the United States. This is an interesting and widely discussed hypothesis. Some strong emotions ride on it, and it is argued that at least it does no harm.

We can do some calculations. Suppose that the elf in question spends one second per house. This isn't quite the usual picture—"Ho, Ho, Ho," and so on—but imagine that he is terribly efficient and very speedy; that would explain why nobody ever sees him very much—only one second per house, after all. With a hundred million houses he has to spend three years just filling stockings. I have assumed he spends no time at all in going from house to house. Even with relativistic reindeer, the time spent in a hundred million houses is three years and not eight hours. This is an example of hypothesis-testing independent of reindeer propulsion mechanisms or debates on the origins of elves. We examine the hypothesis itself, making very straightforward assumptions, and derive a result inconsistent with the hypothesis by many orders of magnitude. We would then suggest that the hypothesis is untenable.

We can make a similar examination, but with greater uncertainty, of the extraterrestrial hypothesis that holds that a wide range of UFOs viewed on the planet Earth are space vehicles from planets of other stars. The report rates, at least in recent years, have been several per day, at the very least. I will not make that assumption. I will make the much more conservative assumption that one such report per year corre-

The rain god, on the frieze of the Temple of the Sun, at San Juan Teotihuacán, Mexico. Photograph by the author.

sponds to a true interstellar visitation. Let's see what this implies.

We have to have some feeling for the number, N, of extant technical civilizations in the Galaxy—that is, civilizations vastly in advance of our own, civilizations that are able, by whatever means, to perform interstellar space flight. (While the means are difficult, they don't enter into this discussion, just as reindeer propulsion mechanisms don't affect our discussion of the Santa Claus hypothesis.)

An attempt has been made to specify explicitly the factors that enter a determination of the number of such technical civilizations in the Galaxy. I will not here run through what numbers have been assigned to the various quantities involved—it's a multiplication of many probabilities, and the likelihood that we can make a good judgment decreases as we proceed down the list. N depends first on the mean rate at which stars are formed in the Galaxy, a number that is known reasonably well. It depends on the number of stars that have planets, which is less well known, but there are some data on that. It depends on the fraction of such planets that are so suitably located with respect to their star that the environment is a feasible one for the origin of life. It depends on the fraction of such otherwise feasible planets on which the origin of life, in fact, occurs. It depends on the fraction of *those* planets on which the origin of life occurs in which, after life has arisen, an intelligent form comes into being. It depends on the fraction of *those* planets in which intelligent forms have arisen that evolve a technical civilization substantially in advance of our own. And it depends on the average lifetime of such a technical civilization.

It is clear that we are rapidly running out of examples as we go farther and farther along. We have many stars, but only one instance of the origin of life, and only a very limited number—some would say only one—of instances of the evolution of intelligent beings and technical civilizations on this planet. And we have no cases whatever to make a judgment on the mean lifetime of a technical civilization. Nevertheless, there is an entertainment that some of us have been engaged in, making our best estimates about these numbers and coming out with a value of N. The result that emerges is that N roughly equals one tenth the average lifetime of a technical civilization in years.

If we put in a number like ten million (10^7) years for the average lifetime of advanced technical civilizations, we come out with a number for such technical civilizations in the Galaxy of about a million (10^6)—that is, a million other stars with planets on which today there are advanced civilizations. This is quite a difficult calculation to do accurately. The choice of ten million years for the average lifetime of a technical civilization is rather optimistic. But let's take these optimistic numbers and see where they lead us.

Let's assume that each of these million technical civilizations launches Q interstellar space vehicles a year, so that 10^6Q interstellar space vehicles are launched per year. Let's assume that there's only one contact made per journey. In the steady-state situation, there are something like 10^6Q arrivals somewhere or other per year. Now, there surely are something like 10^{10} interesting places in the Galaxy to go visit (we have several times 10^{11} stars) and, therefore, an average of $1/10^4=10^{-4}$ arrivals at a given interesting place (let's say a planet) per year. So if only one UFO is to visit the Earth each year, we can calculate what mean launch rate is required at each of these million worlds. The number turns out to be ten thousand launches per year per civilization, and ten billion launches in the Galaxy per year. This seems excessive. Even if we imagine a civilization much more advanced than ours, to launch ten thousand such vehicles for only one to appear here is probably asking too much. And if we were more pessimistic on the lifetime of advanced civilizations, we would require a proportionately larger launch rate. But as the lifetime decreases, the probability that a civilization would develop interstellar flight very likely decreases as well.

There is a related point made by the American physicist Hong-Yee Chiu; he takes more than one UFO arriving at Earth per year, but his argument follows along the same lines as the one I have just presented. He calculates the total mass of metals involved in all of these space vehicles during this history of the Galaxy. The vehicle has to be of some size—it should be bigger than the Apollo capsule, let's say—and we can calculate how much metal is required. It turns out that the total mass of half a million stars has to be processed and all their metals extracted. Or if we extend the argument and assume that only the outer few hundred miles or so of stars like the Sun can be mined by advanced technologies (farther in, it's too hot), we find that two billion such stars must be processed, or about 1 percent of the stars in the Galaxy. This also sounds unlikely.

Now you may say, "Well, that's a very parochial approach; maybe they have plastic spaceships." Yes, I suppose that's possible. But the plastic has to come from somewhere, and plastics vs. metals changes the conclusions very little. This calculation gives some feeling for the magnitude of the task when we are asked to believe that there are routine and frequent interstellar visits to our planet.

What about possible counterarguments? For example, it might be argued that we are the object of special attention—we have just developed all sorts of signs of civilization and high intelligence like nuclear weapons, and maybe, therefore, we are of particular interest to interstellar anthropologists. Perhaps. But we have only signaled the presence of our technical civilization in the past few decades. The news can be only some tens of light-years from us. Also, all the anthropologists in the world do not converge on the Andaman Islands because the fish net has just been invented there. There are a few fish net specialists and a few Andaman specialists; and these guys say, "Well, there's something terrific going on in the Andaman Islands. I've got to spend a year there right away because if I don't go now, I'll miss out." But the pottery experts and the specialists in Australian aborigines don't pack up their bags and leave for the Indian Ocean.

To imagine that there is something absolutely fascinating about what is happening right here is precisely contrary to the idea that there are lots of civilizations around. Because if the latter is true, the development of our sort of civilization must be pretty common. And if we are not pretty common, then there are not going to be many civilizations advanced enough to send visitors.

Even so, is it not possible that the second UFO hypothesis is true—that in historical or recent prehistoric times an extraterrestrial space vehicle made landfall on Earth? There is surely no way in which we can exclude such a contingency. How could we prove it?

A number of popular books have recently been written that allege to demonstrate such a visitation. The arguments are of two sorts, legend and artifact. I broached this subject in the book *Intelligent Life in the Universe*, written with the Soviet astrophysicist I. S. Shklovskii and published in 1966. I examined a typical legend suggestive of contact between our ancestors and an apparent representative of a superior society. The legend, taken from the earliest Sumerian mythology, is important because the Sumerians are the direct cultural antecedents of our own civilization. A superior being was supposed to have taught the Sumerians mathematics, astronomy, agriculture, social and political organization, and written language—all the arts necessary for making the transition from a hunter-gatherer society to the first civilization.

But as provocative as this and similar legends were, I concluded that it was impossible to *demonstrate* extraterrestrial contact from such legends: There are plausible alternative explanations. We can understand why priests might make myths about superior beings who inhabit the skies and give directions to human beings on how to order their affairs. Among other "advantages," such legends permit the priests to control the people.

There is only one category of legend that would be convincing: When information is contained in the legend that could not possibly have been generated by the civilization that created the legend—if, for example, a number transmitted from thousands of years ago as holy turns out to be the nuclear fine structure constant. This would be a case worthy of some considerable attention.

Also convincing would be a certain class of artifact. If an artifact of technology were passed on from an ancient civilization—an artifact that is far beyond the technological capabilities of the originating civilization—we would have an interesting *prima facie* case for extraterrestrial visitation. An example would be an illuminated manuscript, rescued from an Irish monastery, that contains the electronic circuit diagram for a superheterodyne radio receiver. Great care would have to be taken about the provenance of this artifact, just as art collectors are cautious about a newly discovered Raphael. We would make sure that no contemporary Irish prankster was the source of the circuit diagram.

To the best of my knowledge, there are no such legends and no such artifacts. All the ancient artifacts put forward, for example, by Erik von Danniken in his book *Chariots of the Gods?* have a variety of plausible, alternative explanations. Representations of beings with large, elongated heads, alleged to resemble space helmets, could equally well be inelegant artistic renditions, depictions of ceremonial head masks or expressions of rampant hydrocephalia. In fact, the expectation that extraterrestrial astronauts would look precisely like American or Soviet astronauts, down to their space suits and eyeballs, is probably less credible than the idea of a visitation itself. Likewise, the idea expressed by von Danniken and others that ancient astronauts erected airfields, employed rockets, and exploded nuclear weapons on Earth is implausible in the extreme, precisely because we ourselves have just developed this technology. A visitor from space will not be so close to us in time. It is as if, framing such an idea in 1870, we concluded that extraterrestrials use hot-air balloons for space exploration. Far from being too daring, such ideas are stodgy in their unimaginativeness. Most popular accounts of alleged contact with extraterrestrials are strikingly chauvinistic.

An American author named Richard Shaver claims that ordinary rocks, sliced fine, contain a set of still photographs left by an ancient civilization, which can be run as a movie film. Just pick up any rock and slice it fine, he says.

In the great high plain of Nasca in Peru, there is a set of enormous geometrical figures. They are quite difficult to discern when standing among them, but quite discernible from the air. It is easy to see how an early human civilization could have made such figures. But why, it is asked, should such constructs be made except for or by an extraterrestrial civilization? If people believe in the existence of gods in the sky, it is not straining credulity to imagine them making messages to communicate with those gods. The markings may be a kind of collective graphical prayer. But they do not necessarily demonstrate the reality of the intended recipient of the prayer.

There are other cases that seem to be quite convincing at first, such as a perfectly machined steel cube, said to reside in the Salzburg Museum and to have been recovered from geological strata millions of years old, or the receipt of the television call signals of a television station off the air for three years. These cases are almost certainly hoaxes.

There are equally provocative archaeological circumstances that the writers of such sensational books have somehow missed. For example, in the frieze of the great Aztec pyramids at San Juan Teotihuacán, outside Mexico City, there is a repeated figure, described as a rain god, but looking for all the world like an amphibious tracked vehicle with four headlights (see page 201). I do not for a moment believe that such amphibious vehicles were indigenous in Aztec times—among other reasons, because they are too close to what we have today rather than too far from it.

These artifacts are, in fact, psychological projective tests. People can see in them what they wish. There is nothing to prevent anyone from seeing signs of past extraterrestrial visitations all about him. But to a person with an even mildly skeptical mind, the evidence is unconvincing. Because the significance of such a discovery would be so enormous, we must employ the most critical reasoning and the most skeptical attitudes in approaching such data. The data do not pass such tests. Pondering wall paintings, for this purpose, like looking for UFOs, remains an unprofitable investment of terrestrial intelligence—if we are truly interested in the quest for extraterrestrial intelligence.

Cornell University's great radio telescope at Arecibo, Puerto Rico. Courtesy, National Astronomy and Ionosphere Center.

29.

A Search Strategy for Detecting Extraterrestrial Intelligence

Suppose we have arranged a meeting at an unspecified place in New York City with a stranger we have never met and about whom we know nothing—a rather foolish arrangement, but one that is useful for our purposes. We are looking for him, and he is looking for us. What is our search strategy? We probably would not stand for a week on the corner of Seventy-eighth Street and Madison Avenue. Instead, we would recall that there are a number of widely known landmarks in New York City—as well known to the stranger as to us. He knows we know them, we know he knows we know them, and so on. We then shuttle among these landmarks: The Statue of Liberty, the Empire State Building, Grand Central Station, Radio City Music Hall, Lincoln Center, the United Nations, Times Square, and just conceivably, City Hall. We might even indulge ourselves in a few less likely possibilities, such as Yankee Stadium or the Manhattan entrance to the Staten Island Ferry. But there are not an infinite number of possibilities. There are not millions of possibilities; there are only a few dozen possibilities, and in time we can cover them all.

The situation is just the same in the frequency-search strategy for interstellar radio communication. In the absence of any prior contact, how do we know precisely where to search? How do we know which frequency or "station" to tune in on? There are at least millions of possible frequencies with reasonable radio bandpasses. But a civilization interested in communicating with us shares with us a common knowledge about radio astronomy and about our Galaxy. They know, for example, that the most abundant atom in the universe, hydrogen, characteristically emits at a frequency of 1,420 Megahertz. They know we know it. They know we know they know it. And so on. There are a few other abundant interstellar molecules, such as water or ammonia, which have their own characteristic frequencies of emission and absorption. Some of these lie in a region of the galactic radio spectrum where there is less background noise than others. This is also shared information. Students of this problem have come up with a short list of possibly a dozen frequencies that seem to be the obvious ones to examine. It is even conceivable that water-based life will communicate at water frequencies, ammonia-based life at ammonia frequencies, etc.

There appears to be a fair chance that advanced extraterrestrial civilizations are sending radio signals our way, and that we have the technology to receive such signals. How should a search for these signals be organized? Existing radio telescopes, even very small ones, would be adequate for a preliminary search. Indeed, the ongoing search at the Gorky Radiophysical Institute, in the Soviet Union, involves telescopes and instrumentation that are quite modest by contemporary standards.

The amiable and capable president of the Soviet Academy of Sciences, M. V. Keldysh, once told me, with a twinkle in his eye, that "when extraterrestrial intelligence is discovered, then it will become an important scientific problem." A leading American physicist has argued forcefully with me that the best method to search for extraterrestrial intelligence is simply to do ordinary astronomy; if the discovery is to be made, it will be made serendipitously. But it seems to me that we can do something to enhance the likelihood of success in such a search, and that the ordinary pursuit of radio astronomy is not quite the same as an explicit search of certain stars, frequencies, bandpasses, and time constants for extraterrestrial intelligence.

But there are enormous numbers of stars to investigate, and many possible frequencies. A reasonable search program will almost certainly be a very long one. Such a search, using a large telescope full time, should take at least decades, by conservative estimates. The radio observers in such an enterprise, no matter how enthusiastic they may be about the search for extraterrestrial intelligence, would very likely become bored after many years of unsuccessful searching. A radio astronomer, like any other scientist, is interested in working on problems that have a high probability of more immediate results.

The ideal strategy would involve a large telescope that could devote something like half time to the search for extraterrestrial intelligent radio signals and about half time to the study of more conventional radioastronomical objectives, such as planets, radio stars, pulsars, interstellar molecules, and quasars. The difficulty in using several existing radio observatories, each for, say, 1 percent of their time, is that these activities would have to be pursued for many centuries to have a reasonable probability of success. Since the time on existing radio telescopes is mainly spoken for, larger allocations of time seem unlikely.

A wide variety of objects obviously should be examined: G-type stars, like our own; M-type stars, which are older; and exotic objects, which may be black holes or possible manifestations of astroengineering activities. The number of stars and other objects in our own Milky Way Galaxy is about two hundred billion, and the number that we must examine to have a fair chance of detecting such signals seems to be at least millions.

There is an alternative strategy to searching painfully each of millions of stars for the signals from a civilization not much more advanced than our own. We might examine an entire galaxy all at once for signals from civilizations much more advanced than ours (see Chapters 34 and 35). A small radio telescope can point at the nearest spiral galaxy to our own, the great galaxy M31 in the constellation Andromeda, and simultaneously observe some two hundred billion stars. Even if many of these stars were broadcasting with a technology only slightly in advance of our own, we would not pick them up. But if only a few are broadcasting with the power of a much more advanced civilization, we would detect them easily. In addition to examining nearby stars only slightly in advance of us, it therefore makes sense to examine, simultaneously, many stars in neighboring galaxies, only a few of which may have civilizations greatly in advance of our technology.

We have been describing a search for signals beamed in our general direction by civilizations interested in communicating with us. We ourselves are not beaming signals in the direction of some specific other star or stars. If all civilizations listened and none transmitted, we would each reach the erroneous conclusion that the Galaxy was unpopulated, except by ourselves. Accordingly, it has been proposed—as an alternative and much more expensive enterprise—that we also "eavesdrop"; that is, tune in on the signals that a civilization uses for its own purposes, such as domestic radio and television transmission, radar surveillance systems, and the like. A large radio telescope devoting half time to a rigorous search for intelligent extraterrestrial signals beamed our way would cost tens of millions of dollars (or rubles) to construct and operate. An array of large radio telescopes, designed to eavesdrop to a distance of some hundreds of light-years, would cost many billions of dollars.

In addition, the chance of success in eavesdropping may be slight. One hundred years ago we had no domestic radio and television signals leaking out into space. One hundred years from now the development of tight beam transmission by satellites and cable television and new technologies may mean that again no radio and television signals would be leaking into space. It may be that such signals are detectable only for a few hundred years in the multibillion-year history of a planet. The eavesdropping enterprise, in addition to being expensive, may also have a very small probability of success.

The situation we find ourselves in is rather curious. There is at least a fair probability that there are many civilizations beaming signals our way. We have the technology to detect these signals out to immense distances—to the other side of the Galaxy. Except for a few back-burner efforts in the United

States and the Soviet Union, we—that is, mankind—are not carrying out the search for extraterrestrial intelligence. Such an enterprise is sufficiently exciting and, at last, sufficiently respectable that there would be little difficulty in staffing a radio observatory designed for this purpose with devoted, capable, and innovative scientists. The only obstacle appears to be money.

While not small change, some tens of millions of dollars (or rubles) is, nevertheless, an amount of money well within the reach of wealthy individuals and foundations. In fact, there is in astronomy a long and proud history of observatories funded by private individuals and foundations: The Lick Observatory, on Mount Hamilton, California, by Mr. Lick (who wanted to build a pyramid, but settled for an observatory—in the base of which he is buried); the Yerkes Observatory in Williams Bay, Wisconsin, by Mr. Yerkes; the Lowell Observatory in Flagstaff, Arizona, by Mr. Lowell; and the Mount Wilson and Mount Palomar Observatories in Southern California, by a foundation established by Mr. Carnegie. Government money will probably be forthcoming for such an enterprise eventually. After all, it costs about the same as the replacement costs of U.S. aircraft shot down over Vietnam in Christmas week, 1972. But a radio telescope designed for communication with extraterrestrial intelligence and an attached institute of exobiology would make a very fitting personal memorial for someone.

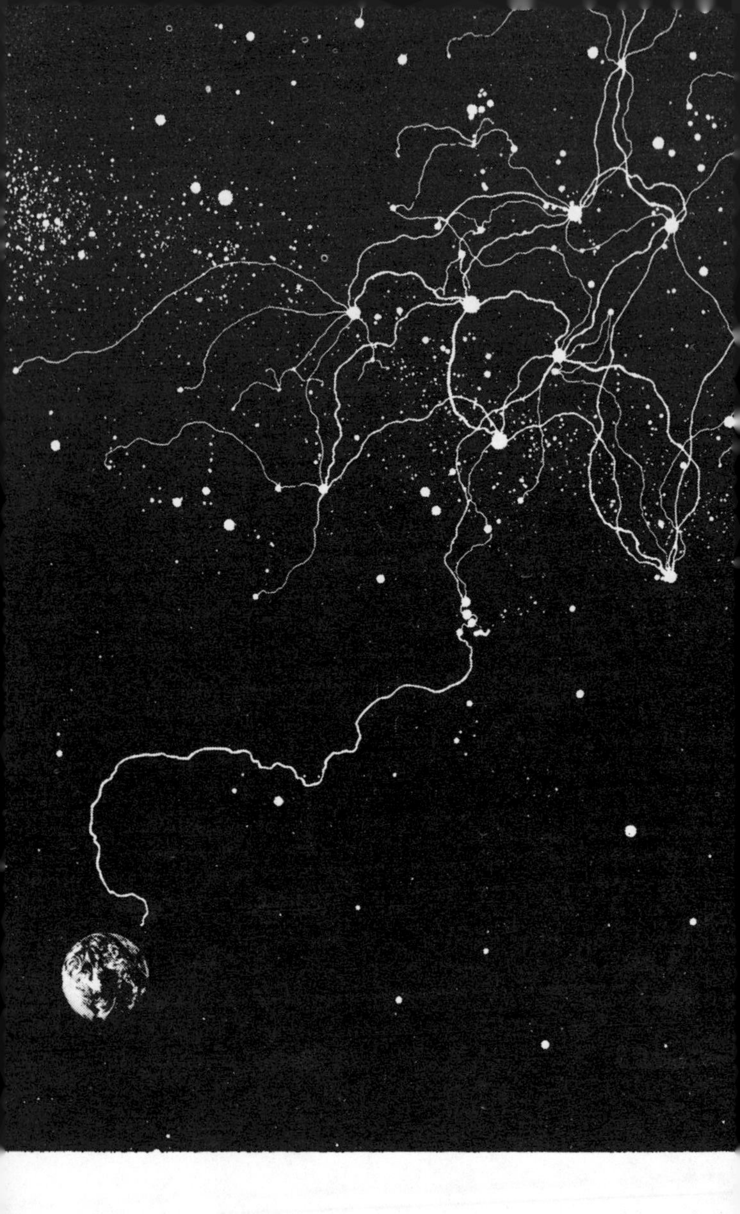

A net of intercommunicating galactic civilizations. By Jon Lomberg.

30.
If We Succeed . . .

In considering the problem of interstellar communication, some people are worried. What if a civilization we come into contact with is more advanced than we?

The history of contact between advanced and backward technological civilizations on Earth is a sorry one. The technically less advanced societies—although they may have superior mathematics or astronomy or poetry or moral precepts—get wiped out. If this is a law of societal natural selection here, why not elsewhere? And in that case, should we not keep quiet?

There are those who predict a dire catastrophe if we broadcast our presence to another star. The extraterrestrials will come and—eat us, or something equally unpleasant. (Actually, if we are especially tasty, they need only sample one of us, determine what sequence of our amino acids makes us appetizing, and then reconstruct the relevant proteins on their own planet. The high freightage makes us economically, if not gastronomically, unappetizing.) The message aboard *Pioneer 10* was criticized by a few because it "gave away" our position in the Galaxy. I very much doubt if we pose any threat to anybody out there. We are the most backward possible civilization able to engage in communication, and the vast spaces between the stars are a kind of natural quarantine, preventing us at any time in the near future from messing around out there.

But, in any case, it is too late. We have already announced our presence. The initial radio broadcasts, starting with Marconi and reaching significant intensity in the 1920s, have leaked through the ionosphere and are expanding at the velocity of light in a spherical wavefront centered around the Earth. And in that wavefront, an advanced technical civilization can pick up the tinny transmissions of Enrico Caruso arias, the Scopes trial, the 1928 election returns, the big jazz bands. These are the harbingers of the cultures of Earth, our first emissaries to the stars.

If there are technical civilizations some fifty light-years out, they will just now be detecting these strange, primitive signals. Even if they are poised to respond instantly with the fastest spaceship possible, it will be at least another fifty years before we hear from them. *Pioneer 10* will take a million years to cover the same distance.

It is too late to be shy and hesitant. We have announced our presence to the cosmos—in a backward and groping and unrepresentative manner, to be sure—but here we are!

The vast distances between the stars imply that there will be no cosmic dialogues by radio transmission. Suppose we receive a signal from a civilization at some likely distance for first contact, such as three hundred light-years. The message says, perhaps, "Hello, you guys; how are you?" Having long been prepared for this moment, we immediately reply, let us say, "Fine, how are you?" The total round-trip communication time would be six hundred years. It's not what you'd call a snappy conversation.

Six hundred years ago, the Black Death stalked Europe, the Ming dynasty was just founded, Charles the Wise sat on the French throne, Gregory XI was Pope, and the Aztecs were hanging those who polluted the water and the air. Six hundred years is a long time on Earth. Interstellar radio communication will not be a dialogue. It will be a monologue. The dumb guys hear from the smart guys, as if the astrologer of Charles the Wise were to receive a message from us.

While the time for radio signals to travel a distance of three hundred light-years is three hundred years, the amount of information that can be conveyed is enormous. In fact, with instrumentation not very much more advanced than our own, essentially all the important insights of our civilization could be transmitted in a few days. It would take three hundred years to get there, but only a few days to be transmitted. But the more lively transmission is in the other direction, from the smarts (them) to the dumbs (us) (see Chapter 31). It is possible that there is a breath-taking repository of galactic knowledge being beamed from several directions at Earth at this moment, advanced text interspersed with primers, so we can learn Galactic, the language of transmission. But we will not hear it if we do not listen for it.

But how could we possibly decode such a message? European scholars spent more than a century in entirely erroneous attempts to decode Egyptian hieroglyphics before the discovery of the Rosetta Stone and the brilliant attack on its translation by Young and Champollion. Some ancient languages, such as the glyphs of Easter Island, the writings of the Mayas, and some varieties of Cretan script, remain completely undecoded to the present time. Yet they were languages of human beings like ourselves, with common biological instincts and encodings, and distant from us in time by only a few hundreds to a few thousands of years. How can we expect that a civilization vastly more advanced than we, and based entirely upon different biological principles, could ever send a message we could understand?

The differences in the two cases are intent and intelligence. The objective of the Easter Island glyphs was not to communicate to twentieth-century scientists. It was to communicate to other Easter Island inhabitants, or, possibly, to the gods. The idea of a code, at least in the usual military intelligence application, is to make a message difficult to read. But the situation we are considering is the opposite. We are considering not cryptography, but anticryptography, the design by a very intelligent civilization of a message so simple that even civilizations as primitive as ours can understand it.

The message will be based upon commonalities between the transmitting and receiving civilizations. Those commonalities are, of course, not any spoken or written language or any common, instinctual encoding in our genetic materials, but rather what we truly share in common—the universe around us, science and mathematics. There are schemes in which mathematical propositions are transmitted, conveying such concepts as addition and equality and negation and then working up to more sophisticated concepts. There are schemes in which radio messages are sent, which, from the number of constituent bits, are clearly pictures; when reconstructed as pictures, they can be clearly understood. The plaque on *Pioneer 10* is an example of a sort of picture which, transmitted as an object on a spacecraft or as a picture by radio transmission, would be reliably understood by an advanced extraterrestrial civilization. Likewise, similar messages, coming our way, will be understood by us, if we have the wit to listen.

Some individuals find the absence of a dialogue distressing —as if meaningful dialogues were commonplace on this planet. Philip Morrison, of the Massachusetts Institute of Technology, has pointed out that such cultural monologues are entirely common in the history of mankind; that, for example, the entire cultural patrimony of classical Greece, which has influenced our civilization in a profound way, has traveled in only one direction in time. We have not sent our wisdom to the Greeks. The Greeks have sent their wisdom to us—on paper and parchment, and not by radio waves, but the principle is the same.

The scientific, logical, cultural, and ethical knowledge to be gained by tuning into galactic transmissions may be, in the long run, the most profound single event in the history of our civilization. There will be information in what we will no longer be able to call the humanities—because our communicants will not be human. There will be a deparochialization of the way we view the cosmos and ourselves. There will be a new perspective on the differences we perceive among ourselves once we grasp the enormous differences that will exist between us and beings elsewhere—beings with whom we have nonetheless a serious commonality of intellectual interest.

But, at the same time, it is not likely to result in discontinuous change. The information may flutter one day into our radio telescopes at a breathtaking rate of information-transfer. The decoding of the message, the understanding of the contents, and the extremely cautious application of what we are taught might take decades or even centuries.

The cultural shock from the content of the message is likely, in the short run, to be small. The main impact will be the receipt of the message itself. Landing of men on the Moon is now considered, at least in the United States, as such a commonplace and relatively uninteresting occurrence that I think we can say the receipt of a message from an extraterrestrial civilization, a message that will take a long time to decode and understand, will not be very much more disorienting to the average man.

Eventually, we may wish to respond.

Why should an advanced society wish to expend the effort to communicate such information to a backward, emerging, novice civilization like our own? I can imagine that they are motivated by benevolence; that during their emerging phases, they were themselves helped along by such messages and that this is a tradition worthy of continuance. There are some science-fiction stories in which the contents of the message are malevolent, in which we receive instructions for the construction of a machine, which we then dutifully build and it then dutifully takes over the Earth. But no one will blindly construct such a machine. No one will implement the instructions contained in an extraterrestrial message until the full theoretical underpinnings and scientific bases of the instructions are well understood. This is one reason why the short-term cultural shock of a message will be small. I do not believe that there is any significant danger from the receipt of such a message, provided the most elementary cautions are adhered to.

It has been suggested that the contents of the initial message received will contain instructions for avoiding our own self-destruction, a possibly common fate of societies shortly after they reach the technical phase. There are certainly enough nuclear weapons on our planet today to destroy every man, woman, and child many times over. It is proposed that advanced extraterrestrial civilizations, motivated either by altruism or through a selfish interest in maintaining a stimulating set of communicants, convey the information for stabilizing societies. I do not know if this is possible; historical differences between organisms and societies with billions of years of independent evolution would be enormous. But it is a possibility not worth ignoring, this feedback hypothesis that the existence of interstellar communication enlarges the number of civilizations and may be the agency of our own survival.

There is another way in which such a feedback process works, even if there are no specific instructions on how to avoid destroying ourselves. There is the matter of time scale. Governments on Earth rarely plan more than five years into the future. Individuals ordinarily make detailed plans for only much shorter times. Even an unsuccessful search for extraterrestrial intelligence, which may take decades or centuries, is a useful example of long-range planning. But think of the consequences of receiving a message that was transmitted three hundred years ago, and a discussion of which will take another six hundred years. Awaiting the answer to our reply requires a continuity of purpose unusual in human institutions. Much of the current ecological catastrophe is due to a grasp of short-term gains and an awesome blindness to long-term disasters. The time scale of interstellar civilizations and discourse with them provides a sense of historical continuity vital for the continuance of our own civilization.

31.
Cables, Drums, and Seashells

In almost all scientific descriptions of contact between Earth and an extraterrestrial civilization, the extraterrestrial civilization is described as advanced.

Why advanced? Why aren't there any primitive civilizations out there, backward fellows poking around, fumbling over interstellar debris, constantly botching things up? Why are we obsessed with advanced civilizations?

The answer is very simple: The primitive ones don't talk to us. (The really smart ones may not talk to us, either, but that is a point I'll come to in a moment.)

Let us consider contact using radio astronomy. Radio astronomy on Earth is a by-product of the Second World War, when there were strong military pressures for the development of radar. Serious radio astronomy emerged only in the 1950s, major radio telescopes only in the 1960s. If we define an advanced civilization as one able to engage in long-distance radio communication using large radio telescopes, there has been an advanced civilization on our planet for only about ten years. Therefore, any civilization ten years less advanced than we cannot talk to us at all.

Even rather optimistic estimates of the rate at which advanced technical civilizations emerge in the Galaxy are lower than one every ten years (see Chapter 28). If this is correct, it means that of all the civilizations in the Galaxy able to communicate by radio, there is none as dumb as we. There may be millions of civilizations less advanced than we, but we have no way to make contact with them: They lack the technology to receive or transmit. Objections that the *Pioneer 10* message may be too difficult for the recipients to decipher ignores the fact that the recipients must be able to acquire this tiny bit of space debris in interstellar space—a task vastly beyond our present capabilities. If they are advanced enough to capture *Pioneer 10* in the dark between the stars, they will, I think, be smart enough to make out its message, which can be read without special hinting by many physicists on as backward a planet as Earth (although, to be sure, those physicists share some genetic and cultural biases and chauvinisms with the authors of the message).

But what about civilizations vastly in advance of our own? The amount of technical progress we have made in the past few hundred years is startling. Not only have entire new technologies developed, but entire new laws of physics and entire new ways of examining the universe have evolved. This intellectual and technological development is continuing. If Earth's civilization survives, the advance of science and technology will also continue.

Civilizations hundreds or thousands or millions of years beyond us should have sciences and technologies so far beyond our present capabilities as to be indistinguishable from magic. It is not that what they can do violates the laws of physics; it is that we will not understand how they are able to use the laws of physics to do what they do.

It is possible that we are so backward and so uninteresting to such civilizations as not to be worthy of contact, or at least of much contact. There may be a few specialists in primitive planetary societies who receive master's or doctor's degrees in studying Earth or listening to our raspy radio and television traffic. There may be amateurs—Boy Scouts, radio hams, and the equivalent—who may be interested in developments on Earth. But a civilization a million years in our future is unlikely, I believe, to be very interested in us. There are all those other civilizations a million years in our future for them to talk to.

Communications between two very advanced civilizations

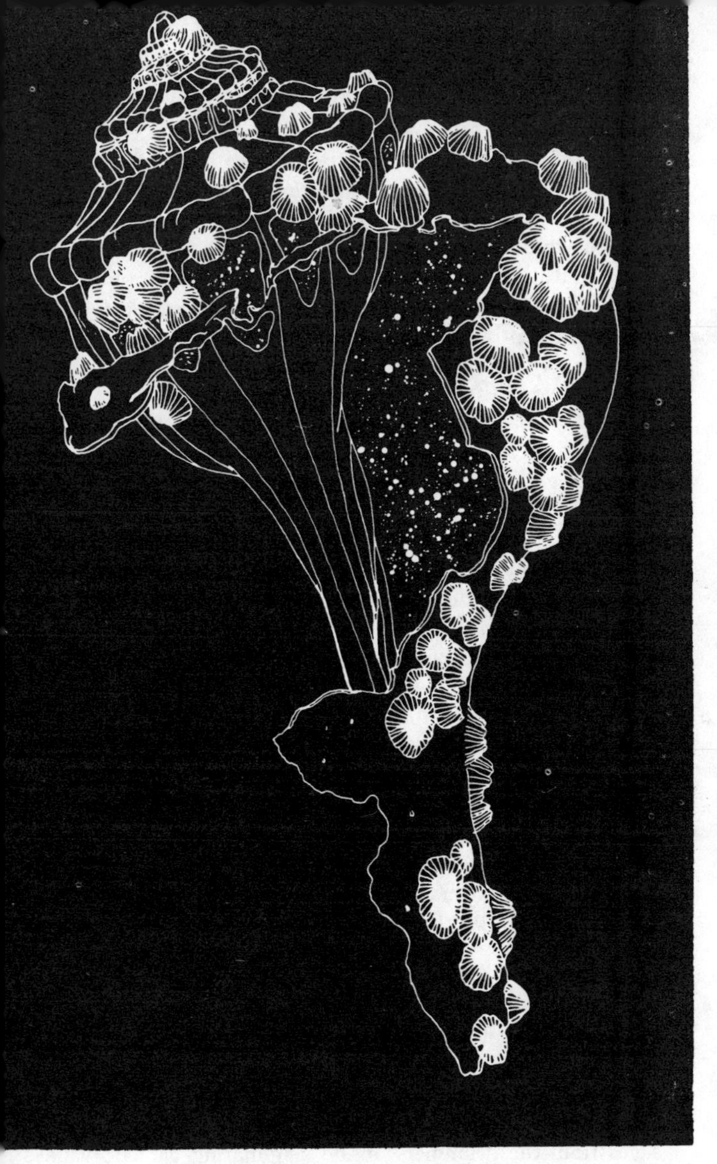

A starshell. By Jon Lomberg.

will likely use a science and a technology inaccessible to us. We therefore have no prospect for tuning in on such communications traffic, either accidentally or on purpose.

We are like the inhabitants of an isolated valley in New Guinea who communicate with societies in neighboring valleys (quite different societies, I might add) by runner and by drum. When asked how a very advanced society will communicate, they might guess by an extremely rapid runner or by an improbably large drum. They might not guess a technology beyond their ken. And yet, all the while, a vast international cable and radio traffic passes over them, around them, and through them.

At this very moment the messages from another civilization may be wafting across space, driven by unimaginably advanced devices, there for us to detect them—if only we knew how. Perhaps the message will come via radio waves to be detected by large radio telescopes. Or perhaps by more arcane devices, the modulation of X-ray stars, gravity waves, neutrinos, tachyons, or transmission channels that no one on Earth will dream of for centuries. Or perhaps the messages are already here, present in some everyday experience that we have not made the right mental effort to recognize. The power of such an advanced civilization is very great. Their messages may lie in quite familiar circumstances.

Consider, for example, seashells. Everyone knows the "sound of the sea" to be heard when putting a seashell to one's ear. It is really the greatly amplified sound of our own blood rushing, we are told. But is this really true? Has this been studied? Has anyone attempted to decode the message being sounded by the seashell? I do not intend this example as literally true, but rather as an allegory. Somewhere on Earth there may be the equivalent of the seashell communications channel. The message from the stars may be here already. But where?

We will listen for the interstellar drums, but we will miss the interstellar cables. We are likely to receive our first messages from the drummers of the neighboring galactic valleys

—from civilizations only somewhat in our future. The civilizations vastly more advanced than we will be, for a long time, remote both in distance and accessibility. At a future time of vigorous interstellar radio traffic, the very advanced civilizations may be, for us, still insubstantial legends.

The night freight to the stars. By Jon Lomberg.

32.
The Night Freight to the Stars

For three generations of human beings there was—as an ever-present, but almost unperceived, part of their lives—a sound that beckoned, a call that pierced the night, carrying the news that there was a way, not so very difficult, to leave Twin Forks, North Dakota, or Apalachicola, Florida, or Brooklyn, New York. It was the wail of the night freight, as haunting and evocative as the cry of the loon. It was a constant reminder that there were vehicles, devices, which, if boarded, could propel you at high velocity out of your little world into a vaster universe of forests and deserts, seacoasts and cities.

Especially in the United States, but perhaps over much of the world, few people today travel by train. There are whole generations growing up which have never heard that siren call. This is the moment of the homogenization of the world, when the diversities of societies are eroding, when a global civilization is emerging. There are no exotic places left on Earth to dream about.

And for this reason there remains an even greater and more poignant need today for a vehicle, a device, to get us somewhere else. Not all of us; only a few—to the deserts of the Moon, the ancient seacoasts of Mars, the forests of the sky. There is something comforting in the idea that one day a few representatives of our little terrestrial village might venture to the great galactic cities.

There are as yet no interstellar trains, no machines to get us to the stars. But one day they may be here. We will have constructed them or we will have attracted them.

And then there will once again be the whistle of the night

freight. Not the antique sort of whistle, for sound does not carry in interplanetary space or in the emptiness between the stars. But there will be something, perhaps the flash from magnetobrehmstrahlung, as the starship approaches the velocity of light. There will be a sign.

Looking out on a clear night from the continent-sized cities and vast game preserves that may be our future on this planet, youngsters will dream that when they are grown, if they are very lucky, they will catch the night freight to the stars.

33. Astroengineering

In a by now much quoted and possibly even apocryphal story, the nuclear physicist Enrico Fermi asked, during a luncheon conversation at Los Alamos in the middle 1940s, "Where are they?" If there are vast numbers of beings more advanced then we, he was musing, why have we seen no sign of them —by a visitation to Earth, for example?

We have discussed this problem in Chapters 27 and 28. But there is another formulation of Fermi's question. A civilization a hundred years in our technological future (assuming present rates of technological growth) would surely be able to communicate by radio, and possibly by other techniques, anywhere in the Galaxy and probably with other galaxies as well. A civilization thousands of years in our technological future will very likely be able to travel physically between the stars, although with the expenditure of considerable time and resources.

But what of civilizations tens of thousands or hundreds of thousands of years in our future, or even farther advanced? There are, after all, stars *billions* of years older than the Sun. The very oldest such stars lack heavy metals; probably their planets are similarly lacking. Such very old stars are inhospitable environments for the development of technological civilizations. But some stars one or two billion years older than the Sun have no such attendant difficulties. It is surely possible that there are at least a few civilizations hundreds of millions or billions of years in our technological future.

With prodigious energy resources, such civilizations should be able to rework the cosmos. We have discussed in Chapter

22 how life on Earth has already altered our planet significantly and how we can envision in the relatively near future making important changes in the environments of the nearby planets.

More major changes are possible in the somewhat more distant future. The mathematician Freeman Dyson, of the Institute for Advanced Study, offers a scheme in which the planet Jupiter is broken down piece by piece, transported to the distance of the Earth from the Sun, and reconstructed into a spherical shell—a swarm of individual fragments revolving about the Sun. The advantage of Dyson's proposal is that all of the sunlight now wasted by not falling upon an inhabited planet could then be gainfully employed; and a population greatly in excess of that which now inhabits the Earth could be maintained. Whether such a vast population is desirable is an important and unsolved question. But what seems clear is that at the present rate of technological progress it will be possible to contruct such a Dyson sphere in perhaps some thousands of years. In that case, other civilizations older than we may have already constructed such spherical swarms.

A Dyson sphere absorbs visible light from the Sun. But it does not continue indefinitely to absorb this light without reradiating; otherwise, the temperature would become impossibly high. The exterior of the Dyson sphere radiates infrared radiation into space. Because of the large dimensions of the sphere, the infrared flux from a Dyson sphere should be detectable over quite sizable distances—with present infrared technology, over distances of hundreds to thousands of light-years. Remarkably enough, large infrared objects of roughly Solar System dimensions and of temperatures less than 1,000 degrees Fahrenheit have been detected in recent years. These, of course, are not necessarily Dyson civilizations. They may be vast dust clouds surrounding stars in the process of formation. But we are beginning to detect objects that are not dissimilar to the artifacts of advanced civilizations.

There are many phenomena in contemporary astronomy that are not understood. Quasars, for example, are one. The reported very high-intensity gravitational waves coming from

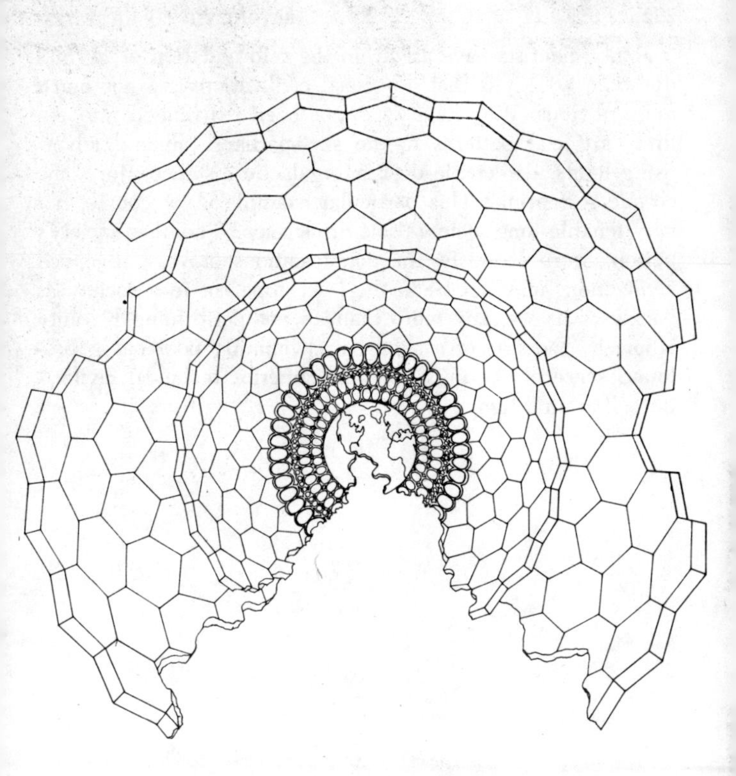

Astroengineering. By Jon Lomberg.

the center of our galaxy are another. The list can be extended considerably. As long as we do not understand these phenomena, we cannot exclude the possibility that they are manifestations of extraterrestrial intelligence. This hardly demonstrates the likelihood of extraterrestrial intelligence, any more than our inability to understand seasonal changes on Mars (Chapter 19) provided strong evidence for vegetation on that planet. As the Soviet astrophysicist I. S. Shklovskii says, "Following the principles of law, we should assume all astronomical phenomena natural until proven otherwise."

Some scientists have asked, in the reformulation of Fermi's question, why it is that advanced civilizations are not much more obvious. Why have stars not been rearranged into entirely artificial patterns in the sky—perhaps blinking advertising lights, detectable over intergalactic distances, for some cosmic soft drink? This particular example is, of course, not very tenable—one society's soft drink may be another society's poison. More seriously, the manifestations of very advanced civilizations may not be in the least apparent to a society as backward as we, any more than an ant performing his anty labors by the side of a suburban swimming pool has a profound sense of the presence of a superior technical civilization all around him.

34. Twenty Questions: A Classification of Cosmic Civilizations

To deal with the possibility of enormously advanced extraterrestrial civilizations, the Soviet astrophysicist N. S. Kardashev has proposed a distinction in terms of the energy available to a civilization for communications purposes.

A Type I civilization is able to muster for communications purposes the equivalent of the entire present power output of the planet Earth—which is now used for heating, electricity, transportation, and so on; a large variety of purposes other than communication with extraterrestrial civilizations. By this definition the Earth is not yet a Type I civilization.

The power usage of our civilization is growing at a rapid rate. The present power output of planet Earth is something like 10^{15} or 10^{16} watts; that is, a million billion to ten million billion watts. The standard exponential notation simply indicates the number of zeros following the 1. For example, 10^{15} means fifteen zeros after the 1. The concept of power in physics is that of an energy expenditure per unit time. One watt is ten million ergs of energy expended per second. All of the power used on the Earth is thus equivalent to lighting up, say, one hundred trillion hundred-watt bulbs. Especially if this energy were put out in the radio part of the spectrum, it might be detected over very sizable distances.

A Type II civilization is able to use for communications purposes a power output equivalent to that of a typical star, about 10^{26} watts. We already see particularly bright stars at optical frequencies in the nearest galaxies. A Type II civilization, putting out in our direction 10^{26} watts in some fairly narrow radio bandpass, could be detectable over vast intergalactic distances. It would be easily detectable, if we used the right search procedures, were there only one such civilization in the nearest spiral galaxy to our own, M31, the great galaxy in the constellation Andromeda. M31 is by no means the largest galaxy. For example, an elliptical galaxy, M87—also known as Virgo A—contains perhaps 10 trillion stars.

Finally, Kardashev imagines a Type III civilization, which would use for communications purposes the energy output of an entire galaxy, roughly 10^{36} watts. A Type III civilization beaming at us could be detected if it were anywhere in the universe. There is no provision for a Type IV civilization, which by definition talks only to itself. There need not be many Type II or Type III civilizations for their presence to be felt once a search for extraterrestrial civilizations is organized in earnest. It may well be that a few Type II or Type III civilizations would be far more readily detectable than a large number of Type I civilizations—if they choose to signal us (see Chapter 31).

The energy gap between a Type I and a Type II civilization or between a Type II and a Type III civilization is enormous—a factor of about ten billion in each instance. It seems useful, if the matter is to be considered seriously, to have a finer degree of discrimination. I would suggest Type 1.0 as a civilization using 10^{16} watts for interstellar communication; Type 1.1, 10^{17} watts; Type 1.2, 10^{18} watts, and so on. Our present civilization would be classed as something like Type 0.7.

But there may be more significant ways to characterize civilizations than by the energy they use for communications purposes. An important criterion of a civilization is the total amount of information that it stores. This information can be described in terms of bits, the number of yes-no statements concerning itself and the universe that such a civilization knows.

An example of this concept is the popular game of "Twenty Questions," as played on Earth. One player imagines an ob-

Type 1.4M Civilization by Jon Lomberg

ject or concept and makes an initial classification of it into animal, vegetable, mineral, or none of these three. To identify the object or concept, the other players then have a total of twenty questions, which can only be answered "Yes" or "No." How much information can be discriminated in this manner?

The initial characterization can be thought of as three yes-no questions: Conceptual or objective? Biological or nonbiological? Plant or animal? If we agree that a particular game of "Twenty Questions" is in pursuit of something alive, we have, in effect, answered three questions already by the time the game begins. The first question divided the universe into two (unequal) pieces. The second question divided one of those pieces into two more, and the third divided one of those pieces into yet two more. At this stage we have divided the universe crudely into $2 \times 2 \times 2 = 2^3 = 8$ pieces. When we have finished with our twenty questions, we have divided the universe into 2^{20} additional (probably unequal) pieces. Now, 2^{10} is 1,024. We can perform such calculations fairly quickly if we approximate 2^{10} by $1{,}000 = 10^3$; therefore, 2^{20} equals $(2^{10})^2$, which approximately equals $(10^3)^2 = 10^6$. The total number of effective questions, twenty-three, has divided the universe into about 2^{23}, or approximately 10^7 pieces or bits. Thus, it is possible for skillful players to win at "Twenty Questions" only if they live in a civilization that has an information content of about 10^7 bits.

But, as I discuss below, our civilization is characterized by perhaps 10^{14} bits. Therefore, skillful players should win at "Twenty Questions" only about 10^7 out of 10^{14} times, or one in 10^7, or one in ten million times. That the game is won more often in practice is because there is an additional rule—usually unstated but well understood: Namely, that the object or concept being named should be one in the general cultural heritage of all the players. But this must mean that 10^7 bits can convey a great deal of information about a civilization, as indeed it can. Philip Morrison has estimated that the total written contribution to our present civilization from classical Greek civilization is only about 10^9 bits. Thus, a one-way message, containing what, by the standards of modern radio astronomy, is a very small number of bits, can contain a very significant amount of new information and can have a powerful influence on a society in the long run.

What is the total number of bits in an English word? In all the books in the world? There are in general English usage twenty-six letters and a sprinkling of punctuation marks. Let us estimate that there are thirty-two such effective "letters." But $32=2^5$; that is, there are something like five bits per letter. If a typical word has four to six letters (for an average of six letters a word, there would have to be a lot of fancy words), there will then be about twenty to thirty bits per word. A typical book—about three hundred words per page and about three hundred pages—would have about a hundred thousand words, or about three million bits. The largest libraries in the world, such as the British Museum, the Bodleian Library at Oxford, the New York Public Library, the Widener Library at Harvard, and the Lenin Library in Moscow, have no more than about ten million volumes. This is about 3×10^{13} bits.

A poor-quality low-resolution photograph may have a million bits in it. A quite complex caricature or cartoon might have only about a thousand bits. On the other hand, a large, high-quality color photograph or painting might have about a billion bits. Let us make allowances for the amount of fundamental information contained in graphics, photography, and art in our civilization, as well as the recorded oral tradition. Let us also try to estimate—this can be done only very crudely —the information we are born with about how to deal with the world. (Human beings are, relative to other animals, born with very little such information—we deal with the world much more in terms of learned rather than inherited or instinctual information.) I estimate, then, that we and our civilization can be very well characterized by something like 10^{14} or 10^{15} bits.

Parenthetically, the ancient Chinese saying that a single

picture is worth ten thousand words (three hundred thousand bits in English; but in Chinese?) is approximately correct—provided that the picture is not too complex.

We can imagine civilizations that have a much greater number of bits characterizing their society than characterizes ours. In general, we would expect a civilization high on the energy scale to be high on the information scale. But this need not necessarily be true. I certainly can imagine societies that are very complex and require many more bits to characterize them than our society requires—but that are not interested in interstellar communication. Characterization of interstellar civilizations requires us to characterize their information content as well.

If we have used numbers to describe energy, we should perhaps use letters to describe information. There are twenty-six letters in the English alphabet. If each corresponds to a factor of ten in the number of bits, there is the possibility of characterizing with the English alphabet a range of information contents over a factor of 10^{26}—a very large range, which seems adequate for our purposes. I propose calling a Type A civilization one at the "Twenty Questions" level, characterized by 10^6 bits. In practice this is an extremely primitive society—more primitive than any human society that we know well—and a good beginning point. The amount of information we have acquired from Greek civilization would characterize that civilization as Type C, although the actual amount of information that characterized Periclean Athens is probably equivalent to Type E or so. By these standards, our contemporary civilization, if characterized by 10^{14} bits of information, corresponds to a Type H civilization.

A combined energy/information characterization of our present global terrestrial society is Type 0.7H. First contact with an extraterrestrial civilization would be, I would guess, with a type such as 1.5J or 1.8K. If there were a galactic civilization of a million worlds, and if each were characterized by a thousand times the information content of our terrestrial civilization, that galactic civilization would be of Type Q. A billion such federated galaxies, with all the information held collectively, would be characterized as a civilization of Type Z.

But as we argue in the next chapter, there is not enough time in the history of the cosmos for such an intergalactic society to have developed. The run of letters from A to Z appears to run the gamut from societies much more primitive than any of Man's to societies more advanced than any that could be.

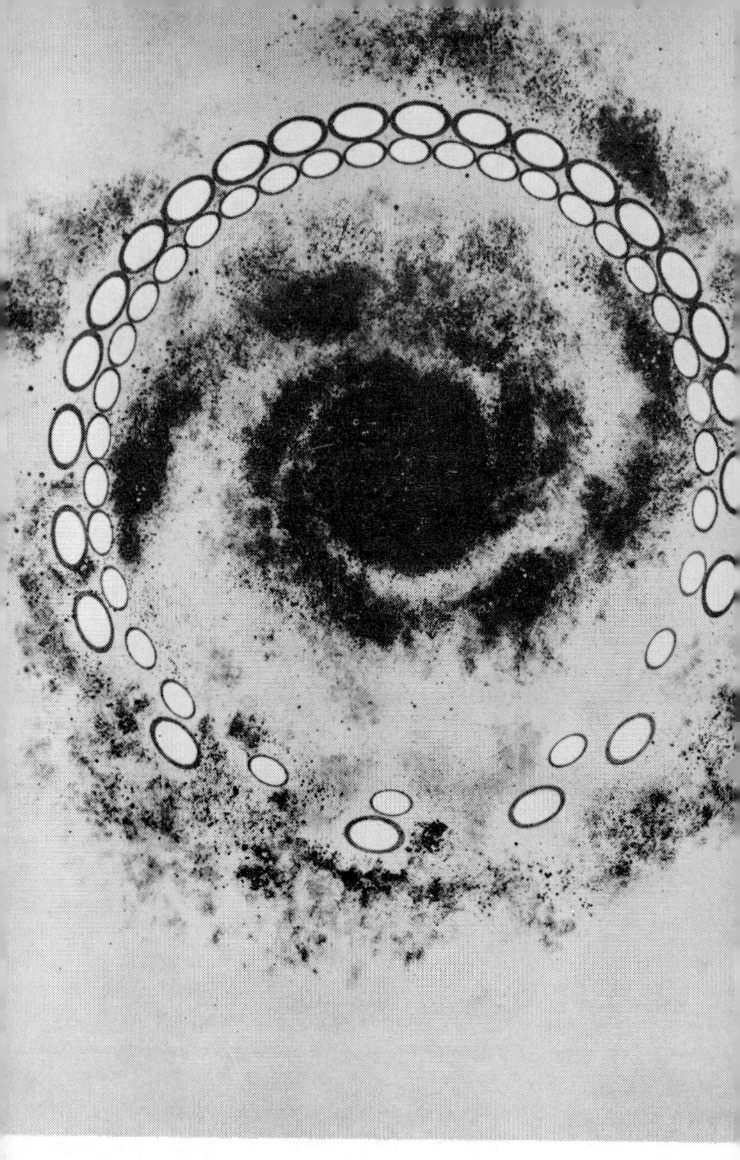

Symbol of a unified galaxy by Jon Lomberg. The galaxy painted is M74 in the constellation Pisces.

35.
Galactic Cultural Exchanges

It is possible to speculate on the very distant future of advanced civilizations. We can imagine such societies in excellent harmony with their environments, their biology, and the vagaries of their politics, so that they enjoy extraordinarily long lifetimes. Communications would long have been established with many other such civilizations. The diffusion of knowledge, techniques, and points of view would occur at the velocity of light. In time, the diverse cultures of the Galaxy, involving a large number of quite different-looking organisms, based on different biochemistries and different initial cultures, would become homogenized—just as the diverse cultures of Earth today are in the process of homogenization.

But such cultural homogenization of the Galaxy will take a long time. One round-trip communication by radio between us and the center of the Milky Way Galaxy requires sixty thousand years. Cultural homogenization of the Galaxy would require many such exchanges, even if each exchange involved very large amounts of information conveyed very efficiently. I find it difficult to believe that fewer than one hundred exchanges between the remotest parts of the Galaxy would be adequate for galactic cultural homogenization.

The minimum lifetime for the homogenization of the Galaxy would thus be many millions of years. The constituent societies must, of course, be stable for comparable periods of time. Such homogenization need not be desirable, but there are still strong and obvious pressures for it to occur, as is also the case on the Earth. If there exists a galactic community of civilizations that truly embraces much of the Milky Way, and if we are right that no information can be transmitted at a velocity faster than light, then most of the members—and all of the founding members—of such a community must be at least millions of years more advanced than we are. For this reason, I think it a great conceit, the idea of the present Earth establishing radio contact and becoming a member of a galactic federation—something like a bluejay or an armadillo applying to the United Nations for member-nation status.

These velocity-of-light limitations on the speed of communication can also be applied to the homogenization of the cultures of different galaxies, after a hypothetical period of millions of years in which the stellar civilizations of a given galaxy achieve a common culture. We can imagine attempts to make contact with such galactic federations in other galaxies.

The nearest spiral galaxies are several million light-years away. This means that a single element of the dialogue—a message and its reply—would take periods of time of several millions to about ten million years. If a hundred such exchanges are required, the time scale for homogenization of a group of nearby galaxies is then of the order of a billion years. The galactic societies would have to be stable and preserve continuity for such periods of time. This would mean that an immensely old civilization within our Galaxy might have strong learned commonalities with similar galactic federations in other members of what astronomers modestly call the "local" group of galaxies.

These homogenization time scales are beginning to reach a point that strains credulity. There are sufficient natural catastrophes and statistical fluctuations in the universe that a stable society—even residing on many different planets simultaneously for more than a billion years—begins to sound unlikely. Also, during these immense periods of time the communicating galactic societies will themselves be evolving; many contacts will be required to maintain homogenization. The galaxies are so distant one from another that they will always retain their cultural individuality.

In any case, all bets are off beyond the local group. To have cultural homogenization with the next such cluster of galaxies like our own, and engage in a hundred message-exchange pairs, would require a time longer than the age of the universe. This is not to exclude long individual messages from one galaxy to another. It may be that enormous amounts of information—about the history of a given galactic federation, for example—may be well known to civilizations in other galaxies. But there will not be enough time for dialogues. At most, one exchange would be possible between the most distant galaxies in the universe. Two exchanges of information at the velocity of light would take more time than there is, according to modern cosmology.

We conclude that there cannot be a strongly cohesive network of communicating, unifying intelligences through the whole universe if (1) such galactic civilizations evolve upward from individual planetary societies and if (2) the velocity of light is indeed a fixed limit on the speed of information transmission, as special relativity requires (i.e., if we ignore such possibilities as using black holes for fast transport: See Chapter 39). Such a universal intelligence is a kind of god that cannot exist.

In a way, St. Augustine and many other thoughtful theologians have come to rather the same conclusion—God must not live from moment to moment, but during all times simultaneously. This is, in a way, the same as saying that special relativity does not apply to Him. But supercivilization gods, perhaps the only ones that this kind of scientific speculation admits, are fundamentally limited. There may be such gods of galaxies, but not of the universe as a whole.

Illustration after H. G. Wells' *The Time Machine*. From Classics Illustrated.

36.
A Passage to Elsewhen

One of the most pervasive and entrancing ideas of science fiction is time travel. In *The Time Machine,* the classic story by H. G. Wells, and in most subsequent renditions, there is a small machine, constructed usually by a solitary scientist in a remote laboratory. One dials the year of interest, steps into the machine, presses a button, and *presto*, here's the past or the future. Among the common devices in time-travel stories are the logical paradoxes that accompany meeting yourself several years ago; killing a lineal antecedent; interfering directly with a major historical event of the past few thousand years; or accidently stepping on a Precambrian butterfly—you are always changing the entire subsequent history of life.

Such logical paradoxes do not occur in stories about travel to the future. Except for the element of nostalgia—the wish we all have to relive or reclaim some elements of the past—a trip forward in time is surely at least as exciting as one backward in time. We know rather much about the past and almost nothing about the future. Travel forward in time has a greater degree of intellectual excitement than the reverse.

There is no question that time travel into the future is possible: We do it all the time merely by aging at the usual rate. But there are other, more interesting possibilities. Everyone has heard about, and now even a fair number of people understand, Einstein's special theory of relativity. It was Einstein's genius to have subjected our usual views of space, time, and simultaneity to a penetrating logical analysis, which could have been performed two centuries earlier. But special relativity required for its discovery a mind divested of the conventional prejudices and the blind adherence to prevailing beliefs—a rare mind in any time.

Some of the consequences of special relativity are counter-intuitive, in the sense that they do not correspond to what everybody knows by observing his surroundings. For example, the special theory says that a measuring rod shrinks in the direction in which it is moving. When jogging, you are thinner in the direction in which you are jogging—and not because of any weight loss. The moment you come to a halt you immediately resume your usual paunch-to-backbone dimension. Similarly, we are more massive when running than when standing still. These statements appear silly only because the magnitude of the effect is too small to be measured at jog velocities. But were we able to jog at some close approximation to the speed of light (186,000 miles per second), these effects would become manifest. In fact, expensive synchrotrons—machines to accelerate charged particles close to the speed of light—take account of such effects, and work only because special relativity happens to be correct. The reason these consequences of special relativity seem counterintuitive is that we are not in the habit of traveling close to the speed of light. It is not that there is anything wrong with common sense; common sense is fine in its place.

There is a third consequence of special relativity, a bizarre effect important only close to the speed of light: The phenomenon called time dilation. Were we to travel close to the speed of light, time, as measured by our wristwatch or by our heartbeat, would pass more slowly than a comparable but stationary clock. Again, this is not an experience of our everyday life, but it is an experience of nuclear particles, which have clocks built into them (their decay times) when they travel close to the speed of light. Time dilation is a measured and authenticated reality of the universe in which we live.

Time dilation implies the possibility of time travel into the future. A space vehicle that could travel arbitrarily close to the speed of light arranges for time, as measured on the space vehicle, to move as slowly as desired. For example, our Galaxy is some sixty thousand light-years in diameter. At the velocity of light, it would take sixty thousand years to cross from one end of the Galaxy to the other. But this time is measured by a stationary observer. A space vehicle able to move close to the speed of light could traverse the Galaxy from one end to the other in less than a human lifetime. With the appropriate vehicle we could circumnavigate the Galaxy and return almost two hundred thousand years later, as measured on Earth. Naturally, our friends and relatives would have changed some in the interval—as would our society and probably even our planet.

According to special relativity, it is even possible to circumnavigate the entire universe within a human lifetime, returning to our planet many billions of years in our future. According to special relativity, there is no prospect of traveling *at* the speed of light, merely very close to it. And there is no possibility in this way of traveling backward in time; we can merely make time slow down, we cannot make it stop or reverse.

The engineering problems involved in the design of space vehicles capable of such velocities are immense. *Pioneer 10*, the fastest man-made object ever to leave the Solar System, is traveling about ten thousand times slower than the speed of light. Time travel into the future is thus not an immediate prospect, but it is a prospect conceivable for an advanced technology on planets of other stars.

There is one further possibility that should be mentioned; it is a much more speculative prospect. At the end of their lifetimes, stars more than about 2.5 times as massive as our Sun undergo a collapse so powerful that no known forces can stop it. The stars develop a pucker in the fabric of space —a "black hole"—into which they disappear. The physics of black holes does not involve Einstein's special theory of relativity; it involves his much more difficult general theory of relativity. The physics of black holes—particularly, rotating black holes—is rather poorly understood at the present time. There is, however, one conjecture that has been made, which cannot be disproved and which is worthy of note: Black holes may be apertures to elsewhen. Were we to plunge down a black hole, we would re-emerge, it is conjectured, in a different part of the universe and in another epoch in time. We do not know whether it is possible to get to this other place in the universe faster down a black hole than by the more usual route. We do not know whether it is possible to travel into the past by plunging down a black hole. The paradoxes that this latter possibility imply could be used to argue against it, but we really do not know.

For all we do know, black holes are the transportation conduits of advanced technological civilizations—conceivably, conduits in time as well as in space. A large number of stars are more than 2.5 times as massive as the Sun; as far as we can tell, they must all become black holes during their relatively rapid evolution.

Black holes may be entrances to Wonderlands. But are there Alices or white rabbits?

37.
Starfolk

I. A FABLE

Once upon a time, about ten or fifteen billion years ago, the universe was without form. There were no galaxies. There were no stars. There were no planets. And there was no life. Darkness was upon the face of the deep. The universe was hydrogen and helium. The explosion of the Big Bang had passed, and the fires of that titanic event—either the creation of the universe or the ashes of a previous incarnation of the universe—were rumbling feebly down the corridors of space.

But the gas of hydrogen and helium was not smoothly distributed. Here and there in the great dark, by accident, somewhat more than the ordinary amount of gas was collected. Such clumps grew imperceptibly at the expense of their surroundings, gravitationally attracting larger and larger amounts of neighboring gas. As such clumps grew in mass, their denser parts, governed by the inexorable laws of gravitation and conservation of angular momentum, contracted and compacted, spinning faster and faster. Within these great rotating balls and pinwheels of gas, smaller fragments of greater density condensed out; these shattered into billions of smaller shrinking gas balls.

Compaction led to violent collisions of the atoms at the centers of the gas balls. The temperatures became so great that electrons were stripped from protons in the constituent hydrogen atoms. Because protons have like positive charges, they ordinarily electrically repel one another. But after a while the temperatures at the centers of the gas balls became so great that the protons collided with extraordinary energy—an energy so great that the barrier of electrical repulsion that surrounds the proton was penetrated. Once penetration occurred, nuclear forces—the forces that hold the nuclei of atoms together—came into play. From the simple hydrogen gas the next atom in complexity, helium, was formed. In the synthesis of one helium atom from four hydrogen atoms there is a small amount of excess energy left over. This energy, trickling out through the gas ball, reached the surface and was radiated into space. The gas ball had turned on. The first star was formed. There was light on the face of the heavens.

The stars evolved over billions of years, slowly turning hydrogen into helium in their deep interiors, converting the slight mass difference into energy, and flooding the skies with light. There were in these times no planets to receive the light, and no life forms to admire the radiance of the heavens.

The conversion of hydrogen into helium could not continue indefinitely. Eventually, in the hot interiors of the stars, where the temperatures were high enough to overcome the forces of electrical repulsion, all the hydrogen was consumed. The fires of the stars were stoked. The pressures in the interiors could no longer support the immense weight of the overlying layers of star. The stars then continued their process of collapse, which had been interrupted by the nuclear fires of a billion years before.

In contracting further, higher temperatures were reached, temperatures so high that helium atoms—the ash of the previous epoch of nuclear reaction—became usable as stellar fuel. More complex nuclear reactions occurred in the insides of the stars—now swollen, distended red giant stars. Helium was converted to carbon, carbon to oxygen and magnesium, oxygen to neon, magnesium to silicon, silicon to sulfur, and upward through the litany of the periodic table of the elements—a massive stellar alchemy. Vast and intricate mazes of nuclear reactions built up some nuclei. Others coalesced to form much more complex nuclei. Still others fragmented or combined with protons to build only slightly more complex nuclei.

The Triffid nebula, a dense cloud of dust and gas out of which bright stars are forming. Courtesy, Steward Observatory, University of Arizona.

But the gravity on the surfaces of red giants is low, because the surfaces have expanded outward from the interiors. The outer layers of red giants are slowly dissipated into interstellar space, enriching the space between the stars in carbon and oxygen and magnesium and iron and all the elements heavier than hydrogen and helium. In some cases, the outer layers of the star were slowly stripped off, like the successive skins of an onion. In other cases, a colossal nuclear explosion rocked the star, propelling at immense velocity into interstellar space most of the outside of the star. Either by leakage or explosion, by dissipation slow or dissipation fast, star-stuff was spewed back to the dark, thin gas from which the stars had come.

But here, later generations of stars were aborning. Again the condensations of gas spun their slow gravitational pirouettes, slowly transmogrifying gas cloud into star. But these new second- and third-generation stars were enriched in heavy elements, the patrimony of their stellar antecedents. Now, as stars were formed, smaller condensations formed near them, condensations far too small to produce nuclear fires and become stars. They were little dense, cold clots of matter, slowly forming out of the rotating cloud, later to be illuminated by the nuclear fires that they themselves could not generate. These unprepossessing clots became the planets: Some giant and gaseous, composed mostly of hydrogen and helium, cold and far from their parent star; others, smaller and warmer, losing the bulk of their hydrogen and helium by a slow trickling away to space, formed a different sort of planet—rocky, metallic, hard-surfaced.

These smaller cosmic debris, congealing and warming, released small quantities of hydrogen-rich gases, trapped in their interiors during the processes of formation. Some gases condensed on the surface, forming the first oceans; other gases remained above the surface, forming the first atmospheres—atmospheres different from the present atmosphere of Earth, atmospheres composed of methane, ammonia, hydrogen sulfide, water, and hydrogen—an unpleasant and unbreathable atmosphere for humans. But this is not yet a story about humans.

Starlight fell on this atmosphere. Storms were driven by the Sun, producing thunder and lightning. Volcanoes erupted, hot lava heating the atmosphere near the surface. These processes broke apart molecules of the primitive atmosphere. But the fragments reassorted into more and more complex molecules, falling into the early oceans, there interacting with each other, falling by chance upon clays, a dizzying process of breakdown, resynthesis, transformation—slowly moving toward molecules of greater and greater complexity, driven by the laws of physics and chemistry. After a time, the oceans achieved the constituency of a warm dilute broth.

Among the innumerable species of complex organic molecules forming and dissipating in this broth there one day arose a molecule able crudely to make copies of itself—a molecule which weakly guided the chemical processes in its vicinity to produce molecules like itself—a template molecule, a blueprint molecule, a self-replicating molecule. This molecule was not very efficient. Its copies were inexact. But soon it gained a significant advantage over the other molecules in the early waters. The molecules that could not copy themselves did not. Those that could, did. The number of copying molecules greatly increased.

As time passed, the copying process became more exact. Other molecules in the waters were reprocessed to form the jigsaw puzzle pieces to fit the copying molecules. A minute and imperceptible statistical advantage of the molecules that could copy themselves was soon transformed by the arithmetic of geometrical progression into the dominant process in the oceans.

More and more elaborate reproductive systems arose. Those systems that copied better produced more copies. Those that copied poorly produced fewer copies. Soon most of the molecules were organized into molecular collectives, into self-replicating systems. It was not that any molecules had the glimmering of an idea or the ghostly passage of a need or want or aspiration; merely, those molecules that copied did, and soon the face of the planet became transformed by the copying process. In time, the seas became full of these molecular collectives, forming, metabolizing, replicating . . . forming, metabolizing, replicating . . . forming, metabolizing, mutating, replicating. . . . Elaborate systems arose, molecular collectives exhibiting behavior, moving to where the replication building blocks were more abundant, avoiding molecular collectives that incorporated their neighbors. Natural selection became a molecular sieve, selecting out those combinations of molecules best suited by chance to further replication.

All the while the building blocks, the foodstuffs, the parts for later copies, were being produced, mainly by sunlight and lightning and thunder—all driven by the nearby star. The nuclear processes in the insides of the stars drove the planetary processes, which led to and sustained life.

As the supply of foodstuffs gradually was exhausted, a new kind of molecular collective arose, one able to produce molecular building blocks internally out of air and water and sunlight. The first animals were joined by the first plants. The animals became parasites upon the plants, as they had been earlier on the stellar manna falling from the skies. The plants slowly changed the composition of the atmosphere; hydrogen was lost to space, ammonia transformed to nitrogen, methane to carbon dioxide. For the first time, oxygen was produced in significant quantities in the atmosphere—oxygen, a deadly posionous gas able to convert all the self-replicating organic molecules back into simple gases like carbon dioxide and water.

But life met this supreme challenge: In some cases by burrowing into environments where oxygen was absent, but—in the most successful variants—by evolving not only to survive the oxygen but to use it in the more efficient metabolism of foodstuffs.

Sex and death evolved—processes that vastly increased the rate of natural selection. Some organisms evolved hard parts, climbed onto, and survived on the land. The pace of production of more complex forms accelerated. Flight evolved. Enormous four-legged beasts thundered across the steaming jungles. Small beasts emerged, born live, instead of in hard-shelled containers filled with replicas of the early oceans. They survived through swiftness and cunning—and increasingly long periods in which their knowledge was not so much preprogrammed in self-replicating molecules as learned from parents and experiences.

All the while, the climate was variable. Slight variations in the output of sunlight, the orbital motion of the planet, clouds, oceans, and polar icecaps produced climatic changes—wiping out whole groups of organisms and causing the exuberant proliferation of other, once insignificant, groups.

And then . . . the Earth grew somewhat cold. The forests retreated. Small arboreal animals climbed down from the trees to seek a livelihood on the savannas. They became upright and tool-using. They communicated by producing compressional waves in the air with their eating and breathing organs. They discovered that organic material would, at a high enough temperature, combine with atmospheric oxygen to produce the stable hot plasma called fire. Postpartum learning was greatly accelerated by social interaction. Communal hunting developed, writing was invented, political structures evolved, superstition and science, religion and technology.

And then one day there came to be a creature whose genetic material was in no major way different from the self-replicating molecular collectives of any of the other organisms on his planet, which he called Earth. But he was able to ponder the mystery of his origins, the strange and tortuous path by which he had emerged from star-stuff. He was the matter of the cosmos, contemplating itself. He considered the problematical and enigmatic question of his future. He called himself Man. He was one of the starfolk. And he longed to return to the stars.

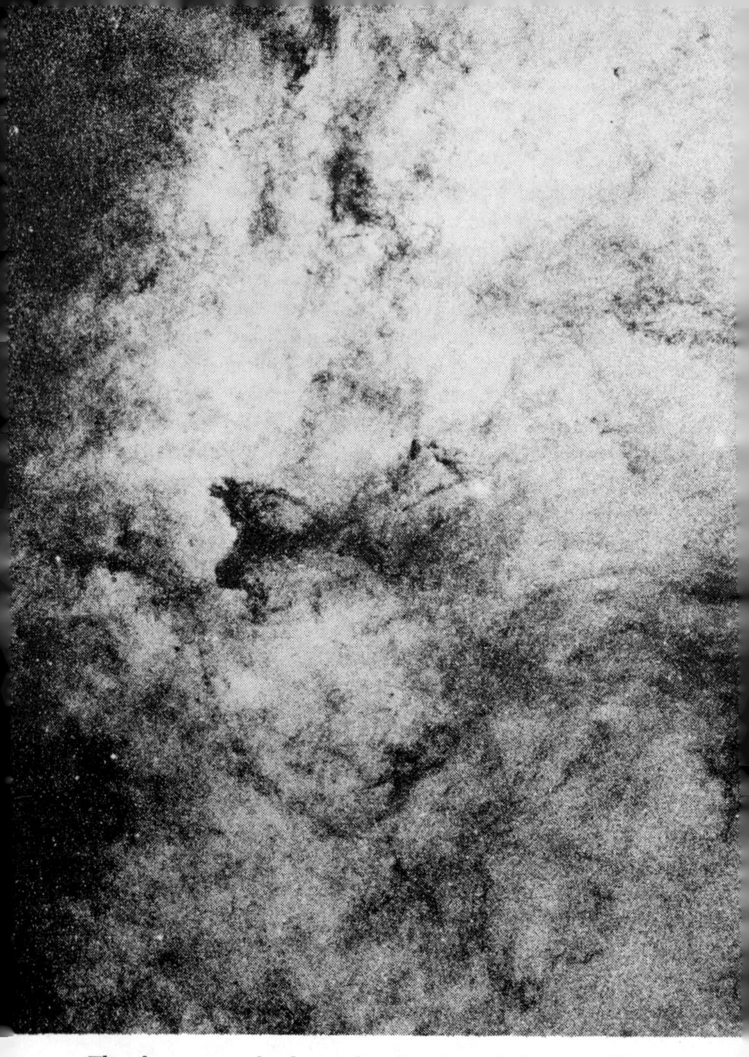

The dense star clouds in the direction of the constellation Sagittarius, toward the center of the Galaxy. The dark lanes are dust clouds where complex organic molecules are being formed. Among these stars, there are some being born and others dying. Innumerable inhabited planets probably circle the stars in this photograph. Courtesy, Hale Observatories.

38.
Starfolk

II. A FUTURE

The story of the preceding chapter is a kind of scientific fable. It is more or less what many modern scientists believe on the basis of available evidence. It is the outline of the emergence of man, a process wending through billions of years of time and driven by gravitation and nuclear physics, by organic chemistry and natural selection. It tells how the matter of which we are made was generated in another place and another time, in the insides of a dying star five billion or more years ago.

There are three aspects of the fable that I find particularly interesting. First, that the universe is put together in such a way as to permit, if not guarantee, the origin of life and the development of complex creatures. It is easy to imagine laws of physics that would not permit appropriate nuclear reactions, or laws of chemistry that would not permit appropriate configurations of molecules to be assembled. But we do not live in such universes. We live in a universe remarkably hospitable to life.

Second, there is in the fable no step unique to our Solar System or to our planet. There are 250 billion suns in our Milky Way Galaxy, and billions of other galaxies in the heavens. Perhaps half of these stars have planets at biologically appropriate distances from the local sun. The initial chemical constituents for the origin of life are the most abundant molecules in the universe. Something like the processes that on Earth led to man must have happened billions of other times in the history of our Galaxy. There must be other starfolk.

The evolutionary details would not be the same, of course. Even if the Earth were started over again and only random forces again operated, nothing like a human being would be produced—because human beings are the end product of an exquisitely complicated evolutionary pathway full of false starts and dead ends and statistical accidents. But we might well expect, if not human beings, organisms functionally not very different from ourselves. Since there are second- and third-generation stars much older than our Sun, there must be, I think, many places in the Galaxy where there are beings far more advanced than we in science and technology, in politics, ethics, poetry, and music.

The third point is the most arresting. It is the intimate connection between stars and life. Our planet was formed from the dregs of star-stuff. The atoms necessary for the origin of life were cooked in the interiors of red giant stars. These atoms were forced together, to form complex organic molecules, by ultraviolet light and thunder and lightning, all produced by the radiation of our neighboring Sun. When the food supply ran short, green-plant photosynthesis developed, driven again by sunlight, the sunlight off which almost all the organisms on Earth, and certainly everyone we know, live out their days.

But this cannot be the end of the fable. Our Sun is only approaching vigorous middle age. It has perhaps another five billion or ten billion years of stellar life ahead of it.

And what of life on Earth and man? They, too, for all we know, may have a future. And if not, there are billions of other stars and probably billions of other inhabited planets in our Galaxy. What is the interaction between stars and life later on?

The death of stars is taking astronomers into unexpected and almost surreal celestial landscapes. One of these is the supernova explosion, the death throes of a star slightly more massive than our Sun. In a brief period of a few weeks to a few months, such an exploding star may become brighter than the rest of the galaxy in which it resides. In supernovae, elements like gold and uranium are generated from iron. The supernovae are the long-sought Philosopher's Stone, converting base metals into precious metals.

Having blown away most of its star-stuff—destined, some of it, to go into later generations of formation of stars and planets and life—the star settles down to a quiet old age, its fires spent, as a white dwarf. A white dwarf is constituted of matter in a state that physicists, with no moral imputation intended, call degenerate. Electrons are stripped off the nuclei of atoms. The protective shields of negative electricity are removed. The nuclei can move much closer together, and a state of extraordinary density results. Typical degenerate matter weighs about a ton per thimbleful. Some white dwarfs, properly considered, are vast stellar crystals able to hold up the weight of the overlying layers of the star. Some white dwarfs are largely carbon. We may speak of a star made of diamond.

But for more massive stars, the white dwarfs—their embers slowly fading, decaying into black dwarfs—are not the end state. Degenerate matter cannot hold up the weight of a more massive star, and another cycle of stellar contraction thus ensues—the matter being crushed together to more and more incredible densities, until some new regime of physics is entered, until some new force surfaces to stop the stellar collapse. There is only one further such force known: It is the nuclear force that holds the nucleus of the atom together. This nuclear force is responsible for the stability of atoms and, therefore, for all of chemistry and biology. It is also responsible for the thermonuclear reactions in the insides of stars that make stars shine and, thereby, in a quite different way, drives planetary biology.

Imagine a star more or less like the Sun, but a little more massive, near the end of its days of converting simple nuclei into more complex nuclei. It produces the last series of complex nuclear reactions it is able to—and then collapses. As its size decreases, it spins more and more rapidly, like a pirouetting ice skater bringing in her arms. Only when the density in its interior becomes comparable to the density of matter inside the atomic nucleus does the collapse stop. It is a simple matter in elementary physics to calculate at what stage the collapse will end. It ends when the star is about a mile across and rotating about ten times a second.

Such an object is a rapidly rotating neutron star. It is, in truth, a giant atomic nucleus a mile across. Neutron star matter is so dense that a speck of it—just barely visible—would weigh a million tons. The Earth would not be able to support it. A piece of neutron star matter, if it could be transported to the Earth without falling apart, would sink effortlessly through the crust, mantle, and core of our planet like a razor blade through warm butter.

Such neutron stars were theoretical constructs, the imaginings of speculative physicists—until the pulsars were discovered. Pulsars are sources of radio emission. Some of them are associated with old supernova explosions. They blink at us as if the beam of some cosmic lighthouse swept by us ten times a second. The details of the emission from pulsars are best understood if they are the fabled neutron stars. Because of the loss of energy to space that we observe, the rotation rate of an isolated neutron star must slowly decline, even though it is a stellar timekeeper of extraordinary accuracy. The observed decay of pulsar periods is just about what is expected from neutron star physics.

The first pulsar to be detected was called, by its discoverers, only half impishly, LGM-1. The LGM stood for "little green men." Was it, they wondered, a beacon from an advanced extraterrestrial civilization? My own view, when I first heard about pulsars, was that they were perfect interstellar navigation beacons, the sorts of markers that an interstellar spacefaring society would want to place throughout the Galaxy for time- and space-fixes for their voyages. There is now little doubt that pulsars are neutron stars. But I would not exclude the possibility that if there are interstellar spacefaring societies, the naturally formed pulsars are used as navigation beacons and for communications purposes.

The state of matter in the inside of such a neutron star is still not understood. We do not know if a surface crust comprising a neutron crystal lattice overlies a core of liquid neutrons. Should the core be solid, starquakes are expected—a shifting of the matter under enormous stress in the interior of the star. Such starquakes should produce a discontinuous change in the period of rotation of the neutron star. Such changes, called "glitches," are observed.

Some were disappointed to learn that pulsars were neutron stars and not interstellar radio communication channels. But pulsars are hardly uninteresting. Indeed, a star more massive than the Sun that fits into a sphere a mile across and rotates ten times every second is, in a certain sense, much more exotic than a civilization slightly more advanced than we, on the planet of another star.

But there is another and much more profound way in which neutron stars and supernova explosions are connected with life. In a supernova explosion, as we have already mentioned, vast quantities of atoms from the surface of the star are ejected at very high velocities into interstellar space. In the case of the neutron star, there is, because of its rapid rotation, a zone, not far from its surface, which is rotating at almost the speed of light. Particles are ejected from that zone at velocities so great that the theory of relativity must be taken into account to describe them. Both supernova explosions and the high-velocity zone surrounding neutron stars must produce cosmic rays—the very fast charged particles (mostly protons, but containing all the other elements as well) that pervade the space between the stars.

Cosmic rays fall on the Earth's atmosphere. The less energetic particles are absorbed by the atmosphere or deflected by Earth's magnetic field. But the more energetic particles, the ones produced by supernovae or neutron stars, penetrate to the surface of the Earth. And here they collide with life. Some cosmic rays penetrate through the genetic material of life forms on the surface of our planet. These random, unpredictable cosmic rays produce changes, mutations, in the hereditary material. Mutations are variations in the blueprints, the hereditary instructions, contained in our self-replicating molecules. Like a fine watch repeatedly hit with a hammer, the functioning of life is unlikely to improve under such random pummelings. But as sometimes happens with watches or bulky television sets, a random pummeling does occasionally improve the functioning. The vast bulk of mutations are harmful, but the small fraction of mutations that are an improvement provide the raw material for evolutionary advance. Life would be at a dead end without mutation. Thus, in yet another way, life on Earth is intimately bound to stellar events. Human beings are here because of the paroxysms in dying stars thousands of light-years away.

The births of stars generate the planetary nurseries of life. The lives of stars provide the energy upon which life depends. The deaths of stars produce the implements for the continued development of life in other parts of the Galaxy. If there are on the planets of dying stars intelligent beings unable to escape their fate, they may at least derive some comfort from the thought that the death of their star, the event that will cause their own extinction, will, nevertheless, provide the means for continued biological advance of the starfolk on a million other worlds.

39.
Starfolk

III. THE COSMIC CHESHIRE CATS

The neutron star is not the most exotic inhabitant of the stellar bestiary. A star larger than three times the mass of our Sun is too big even for nuclear forces to stop its collapse. Once the collapse begins, there is nothing to stop it. The star contracts to one mile across and continues to shrink. The density passes that of the nucleus of an atom, and matter is further crushed together. The gravitational field in the vicinity of such a massive dying star continues to increase. Eventually, the gravity is so strong that not only is matter unable to leave the star, but light is also trapped. A photon traveling at the speed of light away from the star is constrained to follow a curved path and fall back upon the star, just as the Earth's gravity is too strong for any of us to throw a rock to escape velocity. Such stars are too massive for even photons to escape. Consequently, they are dark. They cannot be seen directly, because no light emanates from them. They are present gravitationally, but not optically. They are called "black holes." They are beasts akin to the smile on the Cheshire cat. They are enormous stars that have winked out but are still there.

Black holes are basically theoretical concepts. They may sound no more likely than, and not nearly as charming as, the elves of Middle Earth. But they are probably there. In fact, much of the mass of the Galaxy may reside not in stars we can see, nor in the gas and dust between the stars, but in black holes peppering the Galaxy like the holes in an Emmenthaler cheese.

The first black hole may have been found. Cygnus X-1 is a

rapidly varying source of X-rays, visible light, and radio waves. Its X-ray emission was monitored from NASA's UHURU satellite, launched from an Italian launch complex off the coast of Kenya. All the clues point to Cyg X-1's being a binary star, two stars revolving around each other in a regular and intricate waltz. From the motion of the one star we see, we can deduce the mass of the star we cannot. It turns out to be a massive star, perhaps ten times the mass of our Sun. Such a massive star should ordinarily be very bright. Yet there is no optical hint of its presence. The bright star in Cyg X-1 is revolving about a massive object that is present gravitationally but not optically. It is very likely a black hole a few thousand light-years from Earth.

Black holes may have their uses. What we know about them till now is entirely theoretical, not tested against the skeptical standards of observation. There are some strange possibilities that have been suggested for black holes. Since there is no way to get out of a black hole, it is, in a sense, a separate universe.

In fact, our own universe is very likely itself a vast black hole. We have no knowledge of what lies outside our universe. This is true by definition, but also because of the properties of black holes. Objects that reside in them cannot ordinarily leave them. In a strange sense, our universe may be filled with objects that are not here. They are not separate universes. They do not have the mass of our universe. But in their separateness and their isolation they are autonomous universes.

There is an even more bizarre prospect. In one speculative view (Chapter 36), an object that plunges down a rotating black hole may re-emerge elsewhere and elsewhen—in another place and another time. Black holes may be apertures to distant galaxies and to remote epochs. They may be shortcuts through space and time. If such holes in the fabric of the space-time continuum exist, it is by no means certain that it would ever be possible for an extended object like a spacecraft to use a black hole for travel through space or time. The most serious obstacle would be the tidal force exerted by the

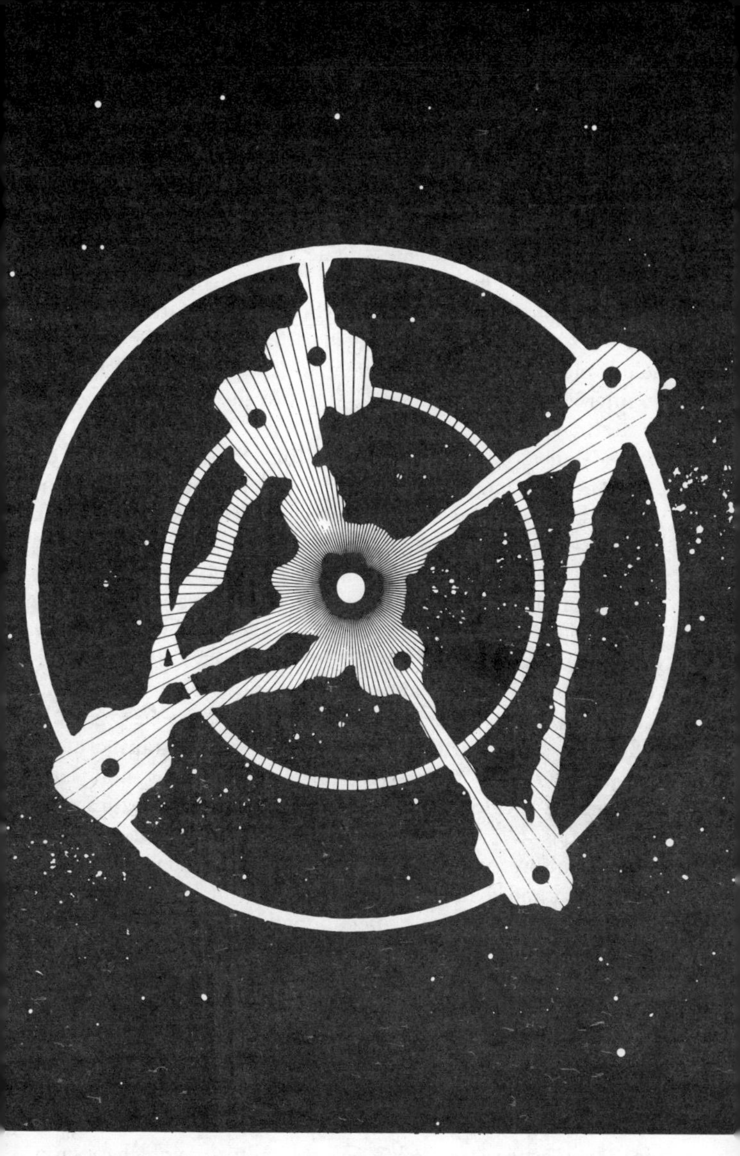

The unification of the cosmos: A conjectural black hole rapid transit system. By Jon Lomberg.

black hole during approach—a force that would tend to pull any extended matter to pieces. And yet it seems to me that a very advanced civilization might cope with the tidal stresses of a black hole.

How many black holes are there in the sky? No one knows at present, but an estimate of one black hole for every hundred stars seems modest by at least some theoretical estimates. I can imagine, although it is the sheerest speculation, a federation of societies in the Galaxy that have established a black hole rapid-transit system. A vehicle is rapidly routed through an interlaced network of black holes to the black hole nearest its destination.

At a typical place in the Galaxy, one hundred stars are encompassed within a volume of radius of about twenty light-years. If we imagine relativistic space vehicles for the short journeys—the local trains or shuttles—it would take only a few years' ship time to get from the black hole to the farthest star of the hundred. One year on board the relativistic shuttle would be occupied accelerating at about 1 g, the acceleration we are familiar with because of the gravity of Earth. After one year at 1 g, we would approach the speed of light. Another year would be spent doing a similar deceleration at 1 g at the end-point of the journey. A galaxy with such a transportation system, a million separately arisen civilizations and large numbers of worlds with colonies, exploratory parties, and work teams—a galaxy where the individuality of the constituent cultures is preserved but a common galactic heritage established and maintained; a galaxy in which the long travel times make trivial contact difficult, and the black hole network makes important contact possible—*that* would be a galaxy of surpassing interest.

I can imagine, in such a galaxy, great civilizations growing up near the black holes, with the planets far from black holes designated as farm worlds, ecological preserves, vacations and resorts, specialty manufacturers, outposts for poets and musicians, and retreats for those who do not cherish big-city life. The discovery of such a galactic culture might happen at any moment—for example, by radio signals sent to the Earth from civilizations on planets of other stars. Or such a discovery might not occur for many centuries, until a lone small vehicle from Earth approaches a nearby black hole and there discovers the usual array of buoys to warn off improperly outfitted spacecraft, and encounters the local immigration officers, among whose duties it is to explain the transportation conventions to newly arrived yokels from emerging civilizations.

The deaths of massive stars may provide the means for transcending the present boundaries of space and time, making all of the universe accessible to life, and—in the last deep sense—unifying the cosmos.

INDEX

INDEX